preface

This book provides detailed revision notes, worked examples and examination questions to support students in their preparation for the new two-tier GCSE Mathematics examinations for the AQA Linear Specification – Foundation Tier.

The book has been designed so that it can be used in conjunction with the companion book *Foundation Mathematics for AQA GCSE* or as a stand-alone revision book for self study and provides full coverage of the new AQA Specifications for the Foundation Tier of entry.

In preparing the text, full account has been made of the requirements for students to be able to use and apply mathematics in written examination papers and be able to solve problems in mathematics both with and without a calculator.

The detailed revision notes, worked examples and examination questions have been organised into 40 self-contained sections which meet the requirements of the National Curriculum and provide efficient coverage of the specifications.

Sections 1 - 11   Number
Sections 12 - 20   Algebra
Sections 21 - 33   Shape, Space and Measures
Sections 34 - 40   Handling Data

At the end of the sections on Number, Algebra, Shape, Space and Measures and Handling Data, section reviews are provided to give further opportunities to consolidate skills.

At the end of the book there is a final examination questions section with a further compilation of exam and exam-style questions, organised for non-calculator and calculator practice, in preparation for the exams.

Also available *Without Answers: (ISBN: 1-405834-90-0)*
The book has been designed so that it can be used in conjunction with the companion book
Foundation Mathematics for AQA GCSE (ISBN: 1-405831-39-1)

contents

Number

Algebra

Student Support Book
with answers

Foundation
MATHEMATICS
for AQA GCSE

Tony Banks and David Alcorn

Causeway
Press

Pearson Education Limited
Edinburgh Gate
Harlow
Essex
CM20 2JE
England

ISBN-13: 978-1-4058-3491-9
ISBN-10: 1-4058-3491-9

Exam questions
Past exam questions, provided by the *Assessment and Qualifications Alliance*, are denoted by the letters AQA. The answers to all questions are entirely the responsibility of the authors/publisher and have neither been provided nor approved by AQA.

Every effort has been made to locate the copyright owners of material used in this book. Any omissions brought to the notice of the publisher are regretted and will be credited in subsequent printings.

Page design
Billy Johnson

Reader
Barbara Alcorn

Artwork
David Alcorn

Cover design
Raven Design

Typesetting by Billy Johnson, San Francisco, California, USA

Printed and bound by Scotprint, Haddington, Scotland

Shape, Space and Measures

Section Review - Shape, Space and Measures

Handling Data

Section Review - Handling Data

Exam Practice

Whole Numbers

- You should be able to read and write numbers expressed in figures and words.

  Eg 1 The number 8543 is written or read as, "eight thousand five hundred and forty-three".

- Be able to order whole numbers.

  Eg 2 Write the numbers 17, 9, 35, 106 and 49 in ascending order.

  9, 17, 35, 49, 106

| Smallest number | ascending order | Largest number |
| Largest number | descending order | Smallest number |

- Be able to recognise the place value of each digit in a number.

  Eg 3 In the number 5384 the digit 8 is worth 80, but in the number 4853 the digit 8 is worth 800.

- Use mental methods to carry out addition and subtraction.

- Know the Multiplication Tables up to 10×10.

- Be able to: multiply whole numbers by 10, 100, 1000, …
  multiply whole numbers by 20, 30, 40, …
  divide whole numbers by 10, 100, 1000, …
  divide whole numbers by 20, 30, 40, …

×	1	2	3	4	5	6	7	8	9	10
1	1	2	3	4	5	6	7	8	9	10
2	2	4	6	8	10	12	14	16	18	20
3	3	6	9	12	15	18	21	24	27	30
4	4	8	12	16	20	24	28	32	36	40
5	5	10	15	20	25	30	35	40	45	50
6	6	12	18	24	30	36	42	48	54	60
7	7	14	21	28	35	42	49	56	63	70
8	8	16	24	32	40	48	56	64	72	80
9	9	18	27	36	45	54	63	72	81	90
10	10	20	30	40	50	60	70	80	90	100

Eg 4 Work out. (a) 75×100
  $= 7500$

  (b) 42×30
  $= 42 \times 10 \times 3$
  $= 420 \times 3$
  $= 1260$

Eg 5 Work out. (a) $460 \div 10$
  $= 46$

  (b) $750 \div 30$
  $= (750 \div 10) \div 3$
  $= 75 \div 3$
  $= 25$

- Use non-calculator methods for addition, subtraction, multiplication and division.

Eg 6 $476 + 254$

```
  4 7 6
+ 2 5 4
-------
  7 3 0
  1 1
```

Eg 7 $374 - 147$

```
  3 ⁶7̷ ¹4
-   1 4 7
-------
    2 2 7
```

Addition and Subtraction
Write the numbers in columns according to place value.
You can use addition to check your subtraction.

Eg 8 324×13

```
    3 2 4
×     1 3
-------
    9 7 2
+ 3 2 4 0
-------
  4 2 1 2
  1 1
```

Long multiplication
Multiply by the units figure, then the tens figure, and so on.
Then add these answers.

Eg 9 $343 \div 7$

```
        4 9
    7)3 4 3
      2 8
      ---
        6 3
        6 3
        ---
          0
```

Long division
÷ (Obtain biggest answer possible.)
Calculate the remainder.
Bring down the next figure and repeat the process until there are no more figures to be brought down.

● Know the order of operations in a calculation.

First	Brackets and Division line
Second	Divide and Multiply
Third	Addition and Subtraction

Eg 10 $4 + 2 \times 6 = 4 + 12 = 16$

Eg 11 $9 \times (7 - 2) + 3 = 9 \times 5 + 3 = 45 + 3 = 48$

Exercise 1   Do not use a calculator for this exercise.

1 (a) Write   five million   in figures.
(b) Write   "six hundred and five thousand two hundred and thirty"   in figures.

2 Write the numbers   85, 9, 23, 117   and   100   in order, largest first.

3 (a) In the number   23 547   the 4 represents 4 tens. What does the 3 represent?
(b) Write the number   23 547   in words.

4 From the list of numbers   2   11   17   29   39   58   71,   write down:
(a) two numbers which have a sum of 60,
(b) two numbers which have a difference of 10.   AQA

5 (a) What must be added to 19 to make 100?
(b) What are the missing values?
    (i) $100 - 65 = \square$ (ii) $12 \times \square = 1200$ (iii) $150 \div \square = 15$

6 Work out.   (a) $769 + 236$ (b) $400 - 209$ (c) $258 - 75$

7 (a) By using each of the digits 8, 5, 2 and 3, write down:
    (i)   the smallest four-digit number,
    (ii)   the largest four-digit odd number.
(b) What is the value of the 5 in the largest number?
(c) What is the value of the 5 in the smallest number?
(d) What is the difference between the two numbers you have written down in part (a)?

8 The chart shows the distances in kilometres between some towns.

Bath			
104	Poole		
153	133	Woking	
362	452	367	Selby

Tony drives from Poole to Bath and then from Bath to Selby.
(a) How far does Tony drive?

Jean drives from Poole to Woking and then from Woking to Selby.
(b) Whose journey is longer?
    How much further is it?

9 Find the missing numbers in these calculations.
(a) $12 + \square = 30$ (b) $14 - \square = 8$ (c) $16 \times \square = 48$ AQA

10 (a) Find the missing numbers in these sentences.
    (i)   When 20 is divided by 6, the remainder is ……
    (ii)   When 20 is divided by ……, the remainder is 4.
(b) Keith says, "When 20 is divided by an odd number, the remainder is always odd."
    Give an example to show that Keith is wrong.   AQA

11 Work out.   (a) 200×60 (b) $40\ 000 \div 80$ (c) 25×7 (d) $45 \div 3$

12 The Lucky Club has 150 members.
Each member pays £15 per year.

Join the Lucky Club
Prizes each month
First prize £25
Second prize £15
Third prize £5

   (a)   How much does the club receive altogether in 1 year?

   (b)   What is the total paid out in prizes in 1 year?

   (c)   How much profit does the club make in 1 year?

AQA

13 Work out.   (a) $500 - 148$ (b) 243×52 (c) $903 \div 43$ AQA

14 A supermarket orders one thousand two hundred tins of beans.
The beans are sold in boxes of twenty-four.
How many boxes of beans are ordered?

15 Work out.   (a) $6 + 4 \times 3$ (b) $96 \div (3 + 5)$ (c) $2 \times (18 - 12) \div 4$

16 John, Paul and Mark go on holiday for 6 days.
Altogether they spend £800.
John spends £350.
Paul spends £40 each day.
On average, how much does Mark spend each day?   AQA

17 Last year Mr Alderton had the following household bills.

Gas	£364	Electricity	£158	Telephone	£187
Water	£244	Insurance	£236	Council Tax	£983

He paid the bills by 12 equal monthly payments.
How much was each monthly payment?

18 Chris is 10 cm taller than Steven. Their heights add up to 310 cm.
How tall is Steven?   AQA

19 A roll of wire is 500 cm long.
From the roll, Debra cuts 3 pieces which each measure 75 cm and 4 pieces which each measure 40 cm.
How much wire is left on the roll?

20 James packs teddy bears into boxes.
He packs 283 teddy bears every hour.
James works 47 hours in one week.
How many teddy bears does James pack in this week?   AQA

21 Mrs Preece is printing an examination for all Year 11 students.
Each examination uses 14 sheets of paper.

   (a)   There are 235 students in Year 11.
      How many sheets of paper does she need?

   (b)   A ream contains 500 sheets of paper.
      How many reams of paper does she need to print all the examinations?   AQA

22   (a)   A travel company takes a party of people to a hockey match at Wembley.
      17 coaches are used.
      Each coach has seats for 46 passengers.
      There are 12 empty seats altogether.
      How many people are in the party?

   (b)   998 football supporters use another travel company to go to a football match at Wembley.
      Each coach has seats for 53 passengers.
      (i)   How many coaches are needed?
      (ii) How many empty seats are there?   AQA

Decimals

What you need to know

- You should be able to write decimals in order by considering place value.

  Eg 1 Write the decimals 4.1, 4.001, 4.15, 4.01, and 4.2 in order, smallest first.
  4.001, 4.01, 4.1, 4.15, 4.2

- Be able to use non-calculator methods to add and subtract decimals.

  Eg 2 2.8 + 0.56

  $$\begin{array}{r} 2.8 \\ + 0.5\,6 \\ \hline 3.3\,6 \\ \hline \end{array}$$

  Eg 3 9.5 − 0.74

  $$\begin{array}{r} {}^{8,\,14,\,1}\llap{9.5}0 \\ - 0.7\,4 \\ \hline 8.7\,6 \\ \hline \end{array}$$

  > **Addition and Subtraction**
  > Keep the decimal points in a vertical column.
  > 9.5 can be written as 9.50.

- You should be able to multiply and divide decimals by powers of 10 (10, 100, 1000, …)

  Eg 4 Work out.
  (a) 6.7×100 = 670
  (b) 0.35×10 = 3.5
  (c) $5.4 \div 10$ = 0.54
  (d) $4.6 \div 100$ = 0.046

- Be able to use non-calculator methods to multiply and divide decimals by other decimals.

  Eg 5 0.43×5.1

  $$\begin{array}{r} 0.4\,3 \quad (2\,\text{d.p.}) \\ \times \quad 5.1 \quad (1\,\text{d.p.}) \\ \hline 4\,3 \leftarrow 43 \times 1 \\ + 2\,1\,5\,0 \leftarrow 43 \times 50 \\ \hline 2.1\,9\,3 \quad (3\,\text{d.p.}) \\ \hline \end{array}$$

  > **Multiplication**
  > Ignore the decimal points and multiply the numbers.
  > Count the total number of decimal places in the question.
  > The answer has the same total number of decimal places.

  Eg 6 $1.64 \div 0.2$

  $$\frac{1.64}{0.2} = \frac{16.4}{2} = 8.2$$

  > **Division**
  > It is easier to divide by a whole number than by a decimal.
  > So, multiply the numerator and denominator by a power of 10 (10, 100, …) to make the dividing number a whole number.

- Be able to use decimal notation for money and other measures.

  > The metric and common imperial units you need to know are given in Section 33.

- Be able to change decimals to fractions.

  Eg 7 (a) $0.2 = \frac{2}{10} = \frac{1}{5}$
  (b) $0.65 = \frac{65}{100} = \frac{13}{20}$
  (c) $0.07 = \frac{7}{100}$

- Be able to carry out a variety of calculations involving decimals.

- Know that when a number is:
  multiplied by a number between 0 and 1 the result will be **smaller** than the original number,
  divided by a number between 0 and 1 the result will be **larger** than the original number.

Exercise 2

Do not use a calculator for questions 1 to 13.

1 Look at this collection of numbers.
Two of these numbers are multiplied together.
Which two numbers will give the smallest answer?

13.5   0.065   0.9   23.0   4.5

3

2 Write the decimals 1.18, 1.80, 1.08, 1.118 in order, smallest first.

3 Work out. (a) 12.08 + 6.51 (b) 6.8 + 4.57 (c) 4.7 − 1.8 (d) 5.0 − 2.36

4 Ann flies from Manchester to Hong Kong.
At Manchester Airport her case is weighed and the scales show 15.7 kg.
In Hong Kong she buys four presents for her family.
They weigh 4 kg, 2.50 kg, 0.75 kg, 3.60 kg.
(a) What is the total weight of these presents in kilograms?
(b) Ann puts the presents in her case when she packs it to fly home.
What does it weigh now?
(c) If her case now weighs more than 20 kg, there is an extra charge.
She has to pay 15 dollars for every kg or part kg over 20 kg.
How much does Ann have to pay?   AQA

5 Cakes cost 27 pence each.
Lubna buys 5 cakes.
She pays with a £10 note.
How much change should she be given?

27p each

6 (a) Multiply 3.2 by 100.
(b) Divide 3.2 by 10.

7 A trader pays £14.80 for 20 melons.
How much does he pay for one melon?   AQA

8 (a) Work out the answer to this sum in your head.
Explain clearly the method you used. 900×0.6
(b) Work out the answer to this sum in your head.
Explain clearly the method you used. $40 \div 0.8$ AQA

9 Two pieces of wood of length 0.75 m and 2.68 m are sawn from a plank 5 m long.
What length of wood is left?

10 (a) Lucy works out 0.2 × 0.4. She gets the answer 0.8.
Explain why her answer must be wrong.
(b) Work out (i) 0.3 × 0.4, (ii) 0.3 × 0.2.

11 Work out. (a) (i) 13.4 × 0.3 (ii) 4.8 × 2.5
(b) (i) 54.4 ÷ 0.4 (ii) 0.294 ÷ 1.2

12 You are given that: 227.5 ÷ 35 = 6.5.
Find the value of: (a) 6.5 × 3.5, (b) 227.5 ÷ 350, (c) 2275 ÷ 0.35. AQA

13 Write as a fraction. (a) 0.3 (b) 0.03 (c) 0.33

14 Rick buys a drink costing £1.35 and some packets of sweets costing 65 pence for each packet.
The total cost is £3.95. How many packets of sweets does Rick buy? AQA

15 David buys 0.6 kg of grapes and 0.5 kg of apples. He pays £1.36 altogether.
The grapes cost £1.45 per kilogram. How much per kilogram are apples? AQA

16 A shopkeeper changed from selling sweets in ounces to selling them in grams.
He used to charge 56p for 4 ounces of sweets.

$$1 \text{ ounce} = 28.4 \text{ grams}$$

How much should he now charge for 125 g of these sweets?   AQA

17 Work out $\dfrac{12.9 \times 7.3}{3.9 + 1.4}$. Write down your full calculator display.

Approximation and Estimation

What you need to know

- How to **round** to the nearest 10, 100, 1000.

  Eg 1   Write 6473 to   (a)   the nearest 10,   (b)   the nearest 100,   (c)   the nearest 1000.
                                  (a)   6470,   (b)   6500,   (c)   6000.

- In real-life problems a rounding must be used which gives a sensible answer.

  Eg 2   Doughnuts are sold in packets of 6. Tessa needs 20 doughnuts for a party.
  How many packets of doughnuts must she buy?

  $20 \div 6 = 3.33\ldots$ This should be rounded up to 4. So, Tessa must buy 4 packets.

- How to approximate using **decimal places**.

  > Write the number using one more decimal place than asked for.
  > Look at the last decimal place and
  > - if the figure is 5 or more round up,
  > - if the figure is less than 5 round down.

  Eg 3   Write the number 3.649 to
        (a)   2 decimal places,
        (b)   1 decimal place.

        (a)   3.65,
        (b)   3.6.

- How to approximate using **significant figures**.

  > Start from the most significant figure and count the required number of figures.
  > Look at the next figure to the right of this and
  > - if the figure is 5 or more round up,
  > - if the figure is less than 5 round down.
  > Add noughts, as necessary, to preserve the place value.

  Eg 4   Write each of these numbers correct to 2 significant figures.
        (a)   365
        (b)   0.0423

        (a)   370
        (b)   0.042

- You should be able to choose a suitable degree of accuracy.

  > The result of a calculation involving measurement should not be given to a greater degree of accuracy than the measurements used in the calculation.

- Be able to use approximations to estimate that the actual answer to a calculation is of the right order of magnitude.

  > Estimation is done by approximating every number in the calculation to one significant figure.
  > The calculation is then done using the approximated values.

  Eg 5   Use approximations to estimate $\dfrac{5.1 \times 57.2}{9.8}$.

  $$\frac{5.1 \times 57.2}{9.8} = \frac{5 \times 60}{10} = 30$$

- Be able to use a calculator to check answers to calculations.

- Be able to recognise limitations on the accuracy of data and measurements.

  Eg 6   Jamie said, "I have 60 friends at my party." This figure is correct to the nearest 10. What is the smallest and largest possible number of friends Jamie had at his party?

  The smallest whole number that rounds to 60 is 55.
  The largest whole number that rounds to 60 is 64.
  So, smallest is 55 friends and largest is 64 friends.

  Eg 7   A man weighs 57 kg, correct to the nearest kilogram.
  What is the minimum weight of the man?
  Minimum weight = 57 kg − 0.5 kg = 56.5 kg.

Do not use a calculator for questions 1 to 19.

1 Write the result shown on the calculator display
(a) to the nearest whole number,
(b) to the nearest ten,
(c) to the nearest hundred.

2 The number of people at a football match was 36 743.
(a) How many people is this to the nearest hundred?
(b) How many people is this to the nearest thousand?

3 A newspaper's headline states: "20 000 people attend concert".
The number in the newspaper is given to the nearest thousand.
What is the smallest possible attendance?

4 Liam wants to calculate $\dfrac{27.89 + 20.17}{3.91}$.
(a) Write each of the numbers in Liam's calculation to the nearest whole number.
(b) Use your numbers from part (a) to estimate the answer to Liam's calculation. AQA

5 The diagram shows the distances between towns *A*, *B* and *C*.

By rounding each of the distances given to the nearest hundred, estimate the distance between *A* and *C*.

6 Wayne is calculating $\dfrac{8961}{1315 + 1692}$.
(a) Write down each of the numbers 8961, 1315 and 1692 to the nearest hundred.
(b) Hence, estimate the value of $\dfrac{8961}{1315 + 1692}$.

7 A queue of cars on the motorway is 3580 metres long.
Cars use an average length of 8.13 metres of roadway.
By rounding each of these numbers to one significant figure, estimate how many cars are in the queue. AQA

8 A snack bar sells coffee at 48 pence per cup.
In one day 838 cups are sold.
By rounding each number to one significant figure, estimate the total amount received from the sale of coffee, giving your answer in pounds. AQA

9

43 × 18 is about 800

(a) Use approximation to show that this is correct.
(b) (i) Show how you could find an estimate for 2019 ÷ 37.
(ii) What is your estimated answer?

10 (a) To estimate 97 × 49, Charlie uses the approximations 100 × 50.
Explain why his estimate will be larger than the actual answer.
(b) To estimate 1067 ÷ 48, Patsy uses the approximations 1000 ÷ 50.
Will her estimate be larger or smaller than the actual answer?
Give a reason for your answer.

11 A college has 6300 students, correct to the nearest hundred.
(a) What is the least possible number of students in this college?
(b) What is the greatest possible number of students in this college? AQA

12 The length of a garden is 50 m, correct to the nearest metre.
What is the minimum length of the garden?

13 Isobella pays for 68 photographs to be developed.
Each photograph costs 34 pence.
Isobella calculates the total cost to be £231.20.
  (a) Which two numbers would you multiply to find a quick estimate of the total cost?
  (b) Use your numbers to show whether Isobella's calculation could be correct.
      Comment on your answer.

AQA

14 Clint has to calculate $\dfrac{414 + 198}{36}$.

He calculates the answer to be 419.5.
By rounding each number to one significant figure, estimate whether his answer is about right.
Show all your working.

15 Melanie needs 200 crackers for an office party.
The crackers are sold in boxes of 12.
How many boxes must she buy?

16 In 2005, Mr Symms drove 8873 kilometres.
His car does 11 kilometres per litre. Petrol costs 89.9 pence per litre.
  (a) By rounding each number to one significant figure, estimate the amount he spent
      on petrol.
  (b) Without any further calculation, explain why this estimate will be larger than the
      actual amount.

17 A running track is 400 m, correct to the nearest metre.
What is the maximum length of the track?

AQA

18 Estimate the value of $\dfrac{407 \times 2.91}{0.611}$.

AQA

19 Sophie spends £8 on a picture. This amount is given correct to the nearest pound.
  (a) Write down the maximum price which Sophie could have paid.
  (b) Write down the minimum price which Sophie could have paid.

AQA

20 Calculate $97.2 \div 6.5$.
Give your answer correct to   (a) two decimal places,   (b) one decimal place.

21 Calculate 78.4×8.7.
Give your answer correct to   (a) two significant figures,   (b) one significant figure.

22 Andrew says, "Answers given to two decimal places are more accurate than answers given
to two significant figures."
Is he right? Explain your answer.

23 Calculate the value of $\dfrac{45.6}{5.3 - 2.78}$.

  (a) Write down your full calculator display.
  (b) Give your answer correct to 3 significant figures.

AQA

24 Use your calculator to evaluate the following. $\dfrac{13.2 + 24.7}{21.3 - 17.2}$

Give your answer correct to one decimal place.

AQA

25 (a) Calculate $\dfrac{89.6 \times 10.3}{19.7 + 9.8}$.

  (b) By using approximations, show that your answer to (a) is about right.
      You **must** show all your working.

AQA

Negative Numbers ● ● ● ● ● ●

What you need to know

- You should be able to use **negative numbers** in context, such as temperature, bank accounts.

- Realise where negative numbers come on a **number line**.

- Be able to put numbers in order (including negative numbers).

  Eg 1 Write the numbers 19, −3, 7, −5 and 0 in order, starting with the smallest.
  $$-5,\ -3,\ 0,\ 7,\ 19$$

- You should be able to add, subtract, multiply and divide with negative numbers.

  Eg 2 Work out.
(a) $-3 + 10$	(b) $-5 - 7$	(c) -4×5	(d) $-12 \div 4$
$= 7$	$= -12$	$= -20$	$= -3$

- Be able to use these rules with negative numbers.

When adding or subtracting:	**When multiplying:**	**When dividing:**
$+ +$ can be replaced by $+$	$+ \times + = +$	$+ \div + = +$
$- -$ can be replaced by $+$	$- \times - = +$	$- \div - = +$
$+ -$ can be replaced by $-$	$+ \times - = -$	$+ \div - = -$
$- +$ can be replaced by $-$	$- \times + = -$	$- \div + = -$

  Eg 3 Work out.
(a) $(-3) + (-2)$	(b) $(-5) - (-8)$	(c) $(-2) \times (-3)$	(d) $(-8) \div (+2)$
$= -3 - 2$	$= -5 + 8$	$= 6$	$= -4$
$= -5$	$= 3$		

Exercise 4   Do not use a calculator for this exercise.

1 What temperatures are shown by these thermometers?

(a)   (b)

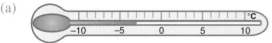   

2 The midday temperatures in three different places on the same day are shown.

Moscow −7°C   Oslo −9°C   Warsaw −5°C

(a) Which place was coldest?
(b) Which place was warmest?

3 The top of a cliff is 125 m above sea level.
The bottom of a lake is 15 m below sea level.
How far is the bottom of the lake below the
top of the cliff?

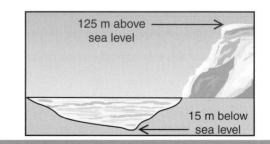

125 m above
sea level

15 m below
sea level

4 Place the following numbers in order of size, starting with the smallest.

$$17 \quad -9 \quad -3 \quad 5 \quad 0 \quad 7$$

5 Work out. (a) $-5 + 10$ (b) $-10 - 5$ (c) -5×10 (d) $-10 \div 5$

6 What number must be placed in the box to complete each of the following?

(a) $-4 + \square = 2$ (b) $-2 - \square = -7$ (c) $-2 - \square = -1$ (d) $\square \times 4 = -12$

7 The table shows the temperature at 9 am on each day of a skiing holiday.

Wednesday	Thursday	Friday	Saturday	Sunday	Monday	Tuesday
$-1°C$	$-2°C$	$-3°C$	$-1°C$	$1°C$	$1°C$	$2°C$

What is the difference in temperature between the warmest day and the coldest day?   AQA

8 In Gdansk, Poland, the temperature on one day in January rose from $-7°F$ to $+22°F$.
By how many degrees did the temperature rise?   AQA

9 Poppy's bank account is overdrawn by £79.
She pays in a cheque and then has a balance of £112.
How much has she paid into her bank account?   AQA

10 The table shows the temperatures recorded at a ski resort one day in February.

Time	0600	1200	1800	2400
Temperature (°C)	-3	3	-2	-6

(a) By how many degrees did the temperature rise between 0600 and 1200?
(b) During which six-hourly period was the maximum drop in temperature recorded?

11 Simon took some chicken pieces out of the freezer.
The temperature of the chicken pieces was $-20°C$.
Two hours later he measured the temperature of the chicken pieces to be $-7°C$.
(a) By how many degrees had the temperature risen?
(b) After another two hours the temperature had risen by the same amount again.
What is the new temperature?   AQA

12 The ice cream is stored at $-25°C$.
How many degrees is this below the required storage temperature?

13 Work out. (a) (i) $(+5) \times (-4)$ (ii) $(-7) \times (-3)$
(b) (i) $(-12) \div (+3)$ (ii) $(-15) \div (-5)$

14 This rule can be used to estimate the temperature in °F for temperatures given in °C.

> Multiply the temperature in °C by 2 and add 30.

Use this rule to estimate $-5°C$ in °F.

15 Work out. (a) $\dfrac{(-2) \times (-5) \times (+6)}{(-3)}$

(b) $(-3) + (-2) \times (+6)$

16 (a) Copy and complete this magic square, so that every row, column and diagonal adds up to 3.
(b) Paul says, "If I multiply each number in the square by -6, the new total for each row, column and diagonal will be -18."
Show **clearly** that this is true for the first row of numbers.

2	–3	4
3	1	
–2		0

AQA

Negative Numbers

What you need to know

- The top number of a fraction is called the **numerator**, the bottom number is called the **denominator**.

- Fractions which are equal are called **equivalent fractions**.

  To write an equivalent fraction:
  Multiply the numerator and denominator by the **same** number.

  Eg 1 $\frac{1}{4} = \frac{1 \times 3}{4 \times 3} = \frac{1 \times 5}{4 \times 5}$

  $\frac{1}{4} = \frac{3}{12} = \frac{5}{20}$

- Fractions can be **simplified** if both the numerator and denominator can be divided by the **same number**. This is sometimes called **cancelling**.

  Eg 2 Write $\frac{20}{28}$ as a fraction in its simplest form.

  $\frac{20}{28} = \frac{20 \div 4}{28 \div 4} = \frac{5}{7}$

  Divide the numerator and denominator by the largest number that divides into both.

- $2\frac{1}{2}$ is an example of a **mixed number**.
  It is a mixture of whole numbers and fractions.

- $\frac{5}{2}$ is an **improper** (or 'top heavy') fraction.

- Fractions must have the **same denominator** before **adding** or **subtracting**.

  Eg 3 Work out.
  (a) $\frac{4}{5} - \frac{1}{2} = \frac{8}{10} - \frac{5}{10} = \frac{3}{10}$
  (b) $2\frac{3}{4} + 1\frac{2}{3} = 2\frac{9}{12} + 1\frac{8}{12} = 3\frac{17}{12} = 4\frac{5}{12}$

  Add (or subtract) the numerators only. When the answer is an improper fraction change it into a mixed number.

- You should be able to multiply and divide fractions.

  Eg 4 Work out.
  (a) $\frac{3}{4} \times \frac{2}{3} = \frac{\cancel{3}^1}{\cancel{4}_2} \times \frac{\cancel{2}^1}{\cancel{3}_1} = \frac{1}{2}$

  The working can be simplified by cancelling.

  (b) $\frac{3}{4} \div \frac{2}{3} = \frac{3}{4} \times \frac{3}{2} = \frac{9}{8} = 1\frac{1}{8}$

  Dividing by $\frac{2}{3}$ is the same as multiplying by $\frac{3}{2}$.

- All fractions can be written as decimals.

  To change a fraction to a decimal divide the **numerator** by the **denominator**.

  Eg 5 Change $\frac{4}{5}$ to a decimal.
  $\frac{4}{5} = 4 \div 5 = 0.8$

- Some decimals have **recurring digits**.
  These are shown by:

  a single dot above a single recurring digit,

  Eg 6 $\frac{2}{3} = 0.6666... = 0.\dot{6}$

  a dot above the first and last digit of a set of recurring digits.

  Eg 7 $\frac{5}{11} = 0.454545... = 0.\dot{4}\dot{5}$

Exercise 5

Do not use a calculator for this exercise.

1 (a) What fraction of this rectangle is shaded?

(b) Copy and shade $\frac{2}{3}$ of this rectangle.

2 Which of the fractions $\frac{2}{5}$, $\frac{3}{8}$, $\frac{4}{12}$, $\frac{5}{15}$ are equivalent?

3 (a) Which of the fractions $\frac{7}{10}$ or $\frac{4}{5}$ is the smaller? Explain why.

(b) Write down a fraction that lies halfway between $\frac{1}{3}$ and $\frac{1}{2}$.

4 Which of the following fractions is nearest to $\frac{1}{2}$?
Show how you decide.

$$\frac{4}{10} \qquad \frac{9}{20} \qquad \frac{14}{30} \qquad \frac{19}{40}$$

AQA

5 Write these fractions in order of size, with the smallest first. $\frac{2}{3}$ $\frac{5}{8}$ $\frac{7}{12}$ $\frac{3}{4}$ AQA

6 Stilton cheese costs £7.20 per kilogram. How much is $\frac{1}{4}$ kg of Stilton cheese?

7 This rule can be used to change kilometres into miles.

Multiply the number of kilometres by $\frac{5}{8}$

Flik cycles 24 kilometres. How many miles is this?

8 Alec's cat eats $\frac{2}{3}$ of a tin of food each day.
What is the least number of tins Alec needs to buy to feed his cat for 7 days? AQA

9 An examination is marked out of 48.
Ashley scored 32 marks.
What fraction of the total did he score?
Give your answer in its simplest form.

10 A garden centre buys 1000 Christmas trees. It sells $\frac{3}{5}$ of them at £8 each.
The remaining trees are then reduced to £5 each and all except 30 are sold.
These 30 trees are thrown away.
How much money does the garden centre get from selling the trees? AQA

11 Elaine buys $\frac{2}{5}$ kg of Edam cheese at £4.80 per kilogram and $\frac{1}{4}$ kg of Cheddar cheese.
She pays £2.83 altogether. How much per kilogram is Cheddar cheese? AQA

12 (a) Change $\frac{1}{6}$ to a decimal. Give the answer correct to 3 d.p.

(b) Write these numbers in order of size, starting with the largest.

$$1.067 \qquad 1.7 \qquad 1.66 \qquad 1\frac{1}{6} \qquad 1.67$$

13 Work out. (a) $\frac{2}{3} \times \frac{1}{4}$ (b) $\frac{2}{3} - \frac{1}{4}$ AQA

14 Income tax and national insurance take $\frac{1}{5}$ of Phillip's pay.
He gives $\frac{2}{5}$ of what he has left to his parents for housekeeping.
What fraction of his pay does Phillip have left for himself?

15 Three-fifths of the people at a party are boys.
Three-quarters of the boys are wearing fancy dress.
What fraction of the people at the party are boys wearing fancy dress?

16 Work out. (a) $3\frac{2}{5} + 2\frac{1}{2}$ (b) $\frac{7}{10} \times \frac{2}{3}$ (c) $\frac{1}{6} \div \frac{1}{9}$

17 Kate is baking two loaves of bread.
One loaf needs $1\frac{1}{4}$ cups of milk. Kate only has $1\frac{2}{3}$ cups of milk.
How much more milk does Kate need? Give your answer as a fraction of a cup. AQA

Working with Number

What you need to know

- **Multiples** of a number are found by multiplying the number by   1,   2,   3,   4,   …

  Eg 1   The multiples of 8 are $1 \times 8 = $ **8**, $2 \times 8 = $ **16**, $3 \times 8 = $ **24**, $4 \times 8 = $ **32**,   …

- **Factors** of a number are found by listing all the products that give the number.

  Eg 2 $1 \times 6 = 6$ and $2 \times 3 = 6$.
  So, the factors of 6 are: 1,   2,   3   and   6.

- The **common factors** of two numbers are the numbers which are factors of **both**.

  Eg 3   Factors of 16 are: 1,   2,   4,   8,   16.
  Factors of 24 are: 1,   2,   3,   4,   6,   8,   12,   24.
  Common factors of 16 and 24 are: 1,   2,   4,   8.

- A **prime number** is a number with only two factors, 1 and the number itself.
  The first few prime numbers are: 2,   3,   5,   7,   11,   13,   17,   19,   …
  The number 1 is not a prime number because it has only one factor.

- The **prime factors** of a number are those factors of the number which are prime numbers.

  Eg 4   The factors of 18 are: 1,   2,   3,   6,   9   and   18.
  The prime factors of 18 are: 2   and   3.

- The **Least Common Multiple** of two numbers is the smallest number that is a multiple of both.

  Eg 5   The Least Common Multiple of 4 and 5 is 20.

- The **Highest Common Factor** of two numbers is the largest number that is a factor of both.

  Eg 6   The Highest Common Factor of 8 and 12 is 4.

- An expression such as $3 \times 3 \times 3 \times 3 \times 3$ can be written in a shorthand way as 3^5.
  This is read as '3 to the power of 5'.
  The number 3 is the **base** of the expression. 5 is the **power**.

- Powers can be used to help write any number as the **product of its prime factors**.

  Eg 7 $72 = 2 \times 2 \times 2 \times 3 \times 3 = 2^3 \times 3^2$

- Numbers raised to the power of 2 are **squared**.
  For example, $3^2 = 3 \times 3 = 9$.
  Squares can be calculated using the $\boxed{x^2}$ button on a calculator.

  > **Square numbers** are whole numbers squared.
  > The first few square numbers are: 1,   4,   9,   16,   25,   36,   …

  The opposite of squaring a number is called finding the **square root**.
  Square roots can be found by using the $\boxed{\sqrt{}}$ button on a calculator.
  The square root of a number can be positive or negative.

  Eg 8   The square root of 9 is $+3$ or -3.

- You should be able to find square roots using a method called **trial and improvement**.
  Work methodically using trials first to the nearest whole number, then to one decimal place, etc.
  Do one trial to one more decimal place than the required accuracy to be sure of your answer.

- Numbers raised to the power of 3 are **cubed**.
  For example, $4^3 = 4 \times 4 \times 4 = 64$.

  > **Cube numbers** are whole numbers cubed.
  > The first few cube numbers are: 1, 8, 27, 64, 125, …

  The opposite of cubing a number is called finding the **cube root**.

  Cube roots can be found by using the $\boxed{\sqrt[3]{}}$ button on a calculator.

- **Powers**

  The squares and cubes of numbers can be worked out on a calculator by using the $\boxed{x^y}$ button.

  The $\boxed{x^y}$ button can be used to calculate the value of a number x raised to the power of y.

  Eg 9 Calculate 2.6^4.
  Enter the sequence: $\boxed{2}$ $\boxed{.}$ $\boxed{6}$ $\boxed{x^y}$ $\boxed{4}$ $\boxed{=}$. So, $2.6^4 = 45.6976$.

- The **reciprocal** of a number is the value obtained when the number is divided into 1.

  Eg 10 The reciprocal of 2 is $\frac{1}{2}$.

  The reciprocal of a number can be found on a calculator by using the $\boxed{\frac{1}{x}}$ button.
  A number times its reciprocal equals 1. Zero has no reciprocal.

- You should be able to simplify calculations involving powers.
  Powers of the same base are **added** when terms are **multiplied**.
  Powers of the same base are **subtracted** when terms are **divided**.

  > In general:
  > $$a^m \times a^n = a^{m+n}$$
  > $$a^m \div a^n = a^{m-n}$$

  Eg 11 (a) $2^3 \times 2^2 = 2^5$ (b) $2^3 \div 2^2 = 2^1 = 2$

You should be able to:

- use the $\boxed{x^2}$, $\boxed{x^y}$, $\boxed{\sqrt{}}$ and $\boxed{\frac{1}{x}}$ buttons on a calculator to solve a variety of problems.

- interpret a calculator display for very large and very small numbers expressed in standard index form.

  Eg 12 $\boxed{1.5 \qquad 10}$ means $1.5 \times 10^{10} = 15\,000\,000\,000$

  $\boxed{6.2 \qquad -05}$ means $6.2 \times 10^{-5} = 0.000\,062$

Exercise 6

Do not use a calculator for questions 1 to 19.

1 Look at these numbers. $\boxed{2 \quad 5 \quad 8 \quad 11 \quad 14 \quad 17 \quad 20}$

(a) Which of these numbers are factors of 10?
(b) Which of these numbers is a multiple of 10?
(c) Which of these numbers are prime numbers?

2 (a) Write down all the factors of 18.
(b) Write down a multiple of 7 between 30 and 40.
(c) Explain why 9 is not a prime number.

3 (a) What is the square of 6?
(b) What is the square root of 100?
(c) What is the cube of 3?
(d) What is the cube root of 8?

4 Work out the value of   (a) 4^2,   (b) 5^3,   (c) 10^4,   (d) $3^2 - 2^3$.

5 James thinks that when you square a number you always get an odd number answer.
Give an example to show that James is wrong.

AQA

6 Jenny says that $2^2 + 3^2 = (2 + 3)^2$. Is she right? Show your working.

7 Look at these numbers. | 2 | 15 | 27 | 36 | 44 | 51 | 64 |

    (a) Which of these numbers is a prime number?
    (b) Which of these numbers is both a square number and a cube number?

8 (a) Work out the value of (i) 5^3,   (ii) $\sqrt{64}$.
    (b) Between which two consecutive whole numbers does $\sqrt{30}$ lie?   AQA

9 (a) Find the square of 4.   (b) Find the square root of 36.
    (c) Find the value of 3×7^2.   (d) Evaluate $(0.1)^2$.   AQA

10 Find the value of (a) $1^2 + 2^2 + 3^2 + 4^2 + 5^2$,   (b) $9^2 \times 10^2$,   (c) $2^3 \times 5^2$.

11 Richard says that $1^3 + 2^3 = 3^3$. Is he right? Show your working.

12 Find the values of (a) $\sqrt{25} + \sqrt{144}$,   (b) $\sqrt{(25 \times 144)}$.   AQA

13 (a) Write 36 as a product of its prime factors.
    (b) Write 45 as a product of its prime factors.
    (c) What is the highest common factor of 36 and 45?
    (d) What is the least common multiple of 36 and 45?

14 A white light flashes every 10 seconds. A red light flashes every 6 seconds.
The two lights flash at the same time.
After how many seconds will the lights next flash at the same time?

15 (a) What is the cube root of 125?
    (b) What is the reciprocal of 4?

16 Find the value of x in each of the following.
    (a) $7^6 \times 7^3 = 7^x$ (b) $7^6 \div 7^3 = 7^x$

17 Simplify fully each of these expressions. Leave your answers in power form.
    (a) $3^2 \times 3^3$ (b) $5^6 \div 5^3$ (c) $\dfrac{2 \times 2^3}{2^2}$

18 (a) Work out 2.5×10^6.
    (b) Write down the number on this calculator display. | 3.7   −0.5 |

19 Which is bigger 2^6 or 3^4? Show all your working.   AQA

20 Use a trial and improvement method to find the square root of 40,
correct to one decimal place. Show your working clearly.   AQA

21 (a) Find the reciprocal of 7, correct to two decimal places.
    (b) Calculate. (i) $3.4 - \frac{1}{1.6}$ (ii) $\sqrt{7.29}$ (iii) $2.4^2 \times \sqrt{1.44}$

22 Calculate $\sqrt{\dfrac{3.9}{(0.6)^3}}$.   AQA

23 (a) Use your calculator to find $3.5^3 + \sqrt{18.4}$. Give all the figures on your calculator.
    (b) Write your answer to 3 significant figures.   AQA

24 (a) Calculate the value of $\sqrt{3.1 + \dfrac{6}{3.1} - \dfrac{9}{3.1^2}}$.
    (b) Show how to check that your answer is of the right order of magnitude.   AQA

25 (a) Calculate. (i) $\sqrt{3.1 - \dfrac{1}{3.1}}$ (ii) $(1.1 + 2.2 + 3.3)^2 - (1.1^2 + 2.2^2 + 3.3^2)$
    (b) Show how you would make a quick estimation of the answer to part (a)(ii).   AQA

SECTION 7 Percentages ● ● ● ● ● ● ● ● ● ● ● ● ● ● ●

What you need to know

- 10% is read as '10 percent'. 'Per cent' means out of 100. 10% means 10 out of 100.

- A percentage can be written as a fraction, 10% can be written as $\frac{10}{100}$.

- To change a decimal or a fraction to a percentage: **multiply by 100**.

  Eg 1 Write as a percentage (a) 0.12 (b) $\frac{8}{25}$

  (a) $0.12 \times 100 = 12\%$ (b) $\frac{8}{25} \times 100 = 32\%$

- To change a percentage to a fraction or a decimal: **divide by 100**.

  Eg 2 Write 18% as (a) a decimal, (b) a fraction.

  (a) $18\% = 18 \div 100 = 0.18$, (b) $18\% = \frac{18}{100} = \frac{9}{50}$.

- How to express one quantity as a percentage of another.

  Eg 3 Write 30p as a percentage of £2.

  $\frac{30}{200} \times 100 = 30 \times 100 \div 200 = 15\%$

  > Write the numbers as a fraction, using the same units.
  > Change the fraction to a percentage.

- You should be able to use percentages to solve a variety of problems.

- Be able to find a percentage of a quantity.

  Eg 4 Find 20% of £64.
  £64 ÷ 100 = £0.64
  £0.64 × 20 = £12.80

  > 1. Divide by 100 to find 1%.
  > 2. Multiply by the percentage to be found.

- Be able to find a percentage increase (or decrease).

  $$\text{Percentage increase} = \frac{\text{actual increase}}{\text{initial value}} \times 100\%$$

  $$\text{Percentage decrease} = \frac{\text{actual decrease}}{\text{initial value}} \times 100\%$$

  Eg 5 Find the percentage loss on a micro-scooter bought for £25 and sold for £18.

  Percentage loss $= \frac{7}{25} \times 100 = 28\%$

Exercise 7   Do not use a calculator for questions 1 to 14.

1 What percentage of these rectangles are shaded?

(a) 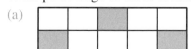   (b)   (c)

2 Copy and complete this table.

Fraction	$\frac{3}{4}$		$\frac{3}{5}$
Decimal	0.75	0.3	
Percentage			

3 Write $\frac{1}{2}$, 0.02 and 20% in order of size, smallest first.

15

4 Work out   (a)   10% of 20 pence,   (b)   25% of 60 kg,   (c)   5% of £900.

5 In an examination, Felicity scored 75% of the marks and Daisy scored $\frac{4}{5}$ of the marks.
Who has the better score? Give a reason for your answer.

6 What is 80 as a percentage of 500?

7 An athletics stadium has 35 000 seats.
4% of the seats are fitted with headphones to help people hear the announcements.
How many headphones are there in the stadium?

8 Jayne is given £50 for her birthday.
She spends 30% of it.
How much of her birthday money does she spend?

9 A large candle costs £4. A medium candle costs 60% of this price.
How much does a medium candle cost?

10 180 college students apply for jobs at a new supermarket.
(a)   70% of the students are given an interview.
    How many students are given an interview?
(b)   54 students are offered jobs.
    What percentage of the students who applied were offered jobs?

11 What is   (a)   60 pence as a percentage of £3,
          (b)   15 seconds as a percentage of 1 minute?

12 The original price of a tennis racket was £35.
What is the sale price of the tennis racket?

○•° SALE °•○
Sports Equipment
20% OFF

13 Mira earns £600 a week. She is given a 5% pay rise.
How much does she now earn a week?

14 A train is travelling at 60 miles per hour.
The train increases its speed to 81 miles per hour.
Calculate the percentage increase in the speed of the train.

15 Find 48% of £9.50.

16 A dress normally costs £35.
The price is reduced by 15% in a sale.
What is the price of the dress in the sale?

17 A snack bar buys packets of sandwiches for £1.60 and sells them for 20% more.
What is the selling price of a packet of sandwiches at the snack bar?

18 In an experiment a spring is extended from 12 cm to 15 cm.
Calculate the percentage increase in the length of the spring.

19 A pogo stick is bought for £12.50 and sold for £8.
What is the percentage loss?

20 A farmer has 200 sheep.
90% of the sheep have lambs.
Of the sheep which have lambs 45% have two lambs.
How many of the sheep have two lambs?

21 At the beginning of the year the value of car A was £3000 and the value of car B was £15 000.
At the end of the year the value of car A is £2400 and the value of car B is £12 150.
Which car has the larger percentage loss? Explain.

Time and Money

What you need to know

- Time can be given using either the **12-hour clock** or the **24-hour clock**.

  Eg 1 (a) 1120 is equivalent to 11.20 am.
  (b) 1645 is equivalent to 4.45 pm.

  > When using the 12-hour clock:
  > times **before** midday are given as am,
  > times **after** midday are given as pm.

- **Timetables** are usually given using the 24-hour clock.

  Eg 2 Some of the rail services from Manchester to Stoke are shown.

Manchester	0925	1115	1215	1415	1555
Stockport	0933	—	1223	—	1603
Stoke	1007	1155	1255	1459	1636

  > Some trains do not stop at every station. This is shown by a dash on the timetable.

  Kath catches the 1555 from Manchester to Stoke.
  (a) How many minutes does the journey take?   (a) 41 minutes.
  (b) What is her arrival time in 12-hour clock time?   (b) 4.36 pm.

- When considering a **best buy**, compare quantities by using the same units.

  Eg 3 Peanut butter is available in small or large jars.
  Small jar: 250 grams for 68 pence.   Large jar: 454 grams for £1.25.
  Which size is the better value for money?

  > Compare the number of grams per penny for each size.

  Small jar: 250 ÷ 68 = 3.67... grams per penny.
  Large jar: 454 ÷ 125 = 3.63... grams per penny.
  The small jar gives more grams per penny and is better value.

- **Value added tax**, or **VAT**, is a tax on some goods and services and is added to the bill.

  Eg 4 A freezer costs £180 + $17\frac{1}{2}$% VAT.

  > $17\frac{1}{2}\% = 17.5\% = \frac{17.5}{100} = 0.175$

  (a) How much is the VAT?   (a) VAT = £180 × 0.175 = £31.50
  (b) What is the total cost of the freezer?   (b) Total cost = £180 + £31.50 = £211.50

- **Exchange rates** are used to show what £1 will buy in foreign currencies.

  Eg 5 Alex buys a painting for 80 euros in France.
  The exchange rate is 1.55 euros to the £.
  What is the cost of the painting in £s?

  1.55 euros = £1   80 euros = 80 ÷ 1.55 = £51.6129...
  The painting cost £51.61, to the nearest penny.

Exercise 8

Do not use a calculator for questions 1 to 5.

1. Greta left home at 2.38 pm and walked for 15 minutes to a bus stop.
   (a) At what time did Greta arrive at the bus stop? Give your answer in 24-hour clock time.

   Her bus arrived at 3.12 pm.
   (b) How long did Greta have to wait for her bus?   AQA

2. Andy buys a box of chocolates which costs £3.57. He pays with a £20 note.
   (a) How much change does he receive?

   This change is given in the smallest number of notes and coins.
   (b) How is the change given?   AQA

3 The times of rail journeys from Guildford to Waterloo are shown.

Guildford	0703	0722	0730	0733	0749	0752
Worplesdon	0708	0727	—	0739	—	0757
Clapham Junction	0752	—	0800	0822	—	—
Waterloo	0800	0815	0808	0830	0823	0844

(a) Karen catches the 0722 from Guildford to Waterloo.
How many minutes does the journey take?

(b) Graham arrives at Worplesdon station at 0715.
What is the time of the next train to Clapham Junction?

4 A train travels from Grantham to London.

(a) The train leaves Grantham at 11.50. It arrives in London at 13.10.
(i) Write these two times in 12-hour clock time.
(ii) How long does the train journey take? Give your answer in hours and minutes.

(b) A family of 2 adults and 1 child travel from Grantham to London.
The adult train fare is £17.68. A child's fare is half the adult's fare.
What is the total cost of their fares?

AQA

5 Reg travels to Ireland. The exchange rate is 1.60 euros to the £.

(a) He changes £40 into euros. How many euros does he receive?

(b) A taxi fare costs 10 euros. What is the cost of the taxi fare in pounds and pence?

6

£980
+ VAT
+ delivery

Total price:
£1197.33

A computer is advertised at £980 + VAT + delivery.
VAT is charged at $17\frac{1}{2}\%$.

(a) What is $17\frac{1}{2}\%$ of £980?

(b) The total price is £1197.33.
What is the charge for delivery?

AQA

7 Nick is on holiday in Spain.
He hires a car at the rates shown.

There are 1.60 euros to £1.

Nick hires the car for 5 days and drives
it for a total of 720 kilometres.
Calculate the total cost of hiring the car.
Give your answer in pounds.

CAR HIRE

Daily rate	54 euros
Free kilometres per day	120
Excess kilometre charge	0.60 euros

8 Toffee is sold in bars of two sizes.
A large bar weighs 450 g and costs £1.69. A small bar weighs 275 g and costs 99p.
Which size of bar is better value for money? You must show all your working.

9 Mrs Tilsed wishes to buy a car priced at £2400.

Two options are available.

Option 1 –
A deposit of 20% of £2400 and 24 monthly payments of £95.

Option 2 –
For a single payment the dealer offers a discount of 5% on £2400.

£2400

How much more does it cost to buy the car if option 1 is chosen rather than option 2?

10 Terry receives a bill for £284 for repairs to his car. VAT at $17\frac{1}{2}\%$ is then added to this amount.
Calculate the total amount which Terry pays.

AQA

What you need to know

- **Hourly pay** is paid at a **basic rate** for a fixed number of hours.
  Overtime pay is usually paid at a higher rate such as time and a half, which means each hour's work is worth 1.5 times the basic rate.

  Eg 1 Alexis is paid £7.20 per hour for a basic 35-hour week.
  Overtime is paid at time and a half.
  Last week she worked 38 hours. How much was Alexis paid last week?

  Basic pay $= £7.20 \times 35$ $= £252$
  Overtime pay $= 1.5 \times £7.20 \times 3 = £\ 32.40$

  Total pay $= £252 + £32.40$ $= £284.40$

- Everyone is allowed to earn some money which is not taxed. This is called a **tax allowance**.

- Tax is only paid on income earned in excess of the tax allowance. This is called **taxable income**.

  Eg 2 Tom earns £6080 per year. His tax allowance is £4895 per year and he pays tax at 10p in the £ on his taxable income. Find how much income tax Tom pays per year.

  Taxable income $= £6080 - £4895 = £1185$
  Income tax payable $= £1185 \times 0.10$ $= £118.50$

  > First find the taxable income, then multiply taxable income by rate in £.

- Gas, electricity and telephone bills are paid **quarterly**.
  Some bills consist of a standing charge plus a charge for the amount used.

- Money invested in a savings account at a bank or building society earns **interest**.
  Simple Interest is when the interest is paid out each year and not added to your account.
  Simple Interest = Amount invested × Time in years × Rate of interest per year.

  Eg 3 Find the Simple Interest paid on £600 invested at 5% for 6 months.

  Simple Interest $= \frac{600}{1} \times \frac{6}{12} \times \frac{5}{100} = £15$

  > 6 months $= \frac{6}{12}$ years.

- You should be able to work out a variety of problems involving personal finance.

Exercise 9

Do not use a calculator for questions 1 to 6.

1. Jenny worked $2\frac{1}{2}$ hours at £6.50 per hour. How much did she earn?

2. Amrit pays his council tax by 10 instalments.
   His first instalment is £193.25 and the other 9 instalments are £187 each.
   How much is his total council tax?

3. Joe insures his house for £180 000 and its contents for £14 000.
   The annual premiums for insurance are:
           House: 23p per annum for every £100 of cover.
           Contents: £1.30 per annum for every £100 of cover.
   What is the total cost of Joe's insurance?

   AQA

4. Last year Harry paid the following gas bills.

£196.40	£62.87	£46.55	£183.06

   This year he will pay his gas bills by 12 equal monthly payments.
   Use last year's gas bills to calculate his monthly payments.

5 Kate is paid £8.17 per hour for the 38 hours she has worked in a week.
Use suitable approximations to estimate Kate's pay for that week.
You must show all your working. AQA

6 Karen has an annual income of £6095. She has a tax allowance of £4895.
(a) Calculate Karen's taxable income.

She pays tax at the rate of 10p in the £ on her taxable income.
(b) How much tax does Karen pay per year? AQA

7 Angela is paid £6.90 per hour for a basic 35-hour week. Overtime is paid at time and a half.
One week Angela worked $37\frac{1}{2}$ hours. How much did Angela earn that week?

8 Sharon is a holiday tour representative in Ibiza.
She is paid a basic wage of £600 per month. She also receives a commission of 5% of
the cost of the excursions which she sells to her holidaymakers.
In **each** of the 4 weeks of July, she sells £1200 of excursions.
What is Sharon's total pay for July? AQA

9 Andy has a part-time job. He is paid £46.50 for working from 8.00 am to 3.30 pm.
How much is this per hour? AQA

10 The cost of a new house is made up of two parts.

Cost of land	£75 000
Building cost	£130 per square foot

A house has an area of 1700 square feet. A builder buys the land and builds a house.
He sells it for £350 000. How much profit is made by the builder? AQA

11 Mr Patel has a faulty shower.
He calls out a plumber who replaces
some parts costing £18.60.
Copy and complete the bill.

Andy's Plumbers

Fixed call charge	£27.50
1¼ hours at £18 per hour	
Parts	£18.60
Total, before VAT	
Vat at 17½%	
Total due	

AQA

12 A monthly income bond pays 6% simple interest per year.
Interest is paid monthly. John invests £15 000.
How much interest is he paid each month?

13 Leroy earns £13 880 per year.
He has a tax allowance of £4895 and pays tax at the rate of 10p in the £ on the first £2090
of his taxable income and 22p in the £ on the remainder.
How much income tax does he pay each year?

14 An electricity bill is made up of two parts.

a fixed charge of £11.64, and
a charge of 7.34p for each unit of electricity used.

VAT at 5% is added to the total.
(a) Hannah uses 743 units of electricity.
Calculate her electricity bill.
(b) Simon receives an electricity bill for £101.12 excluding VAT.
Calculate how many units of electricity Simon has used. AQA

Ratio and Proportion ● ● ● ● ● ● ● ●

What you need to know

- The ratio 3 : 2 is read '3 to 2'.

- A ratio is used only to **compare** quantities.
  A ratio does not give information about the exact values of quantities being compared.

- Different forms of the **same ratio**, such as 2 : 1 and 6 : 3, are called **equivalent ratios**.

- In its **simplest form**, a ratio contains whole numbers which have no common factor other than 1.

  Eg 1 Write £2.40 : 40p in its simplest form.
  £2.40 : 40p = 240p : 40p
  $\qquad\qquad$ = 240 : 40
  $\qquad\qquad$ = 6 : 1

  > All quantities in a ratio must be in the **same units** before the ratio can be simplified.

- You should be able to solve a variety of problems involving ratio.

  Eg 2 The ratio of bats to balls in a box is 2 : 3.
  There are 12 bats in the box.
  How many balls are there?

  $12 \div 2 = 6$
  $2 \times 6 : 3 \times 6 = 12 : 18$
  There are 18 balls in the box.

  > For every 2 bats there are 3 balls. To find an equivalent ratio to 2 : 3, in which the first number is 12, multiply each number in the ratio by 6.

  Eg 3 A wall costs £600 to build.
  The costs of materials to labour are in the ratio 1 : 4.
  What is the cost of labour?

  $1 + 4 = 5$
  $£600 \div 5 = £120$
  Cost of labour $= £120 \times 4 = £480$

  > The numbers in the ratio add to 5. For every £5 of the total cost, £1 pays for materials and £4 pays for labour. So, **divide** by 5 and then **multiply** by 4.

- When two different quantities are always in the **same ratio** the two quantities are in **direct proportion**.

  Eg 4 20 litres of petrol cost £14.
  Find the cost of 25 litres of petrol.

  20 litres cost £14
  1 litre costs £14 ÷ 20 = £0.70
  25 litres cost £0.70 × 25 = £17.50

  > This is sometimes called the **unitary method**.
  > **Divide** by 20 to find the cost of 1 litre.
  > **Multiply** by 25 to find the cost of 25 litres.

Exercise 10

Do not use a calculator for questions 1 to 5.

1 Write these ratios in their simplest form.
(a) 2 : 6 $\qquad$ (b) 8 : 4 $\qquad$ (c) 6 : 9

2 Rhys draws a plan of his classroom floor. The classroom measures 15 m by 20 m.
He draws the plan to a scale of 1 cm to 5 m.
What are the measurements of the classroom floor on the plan?

3 A toy box contains large bricks and small bricks in the ratio 1 : 4.
The box contains 40 bricks. How many large bricks are in the box?

4 To make mortar a builder mixes sand and cement in the ratio 3 : 1.
The builder uses 2.5 kg of cement.
How much sand does he use?

5 In a drama club the ratio of boys to girls is 1 : 3.
  (a) What fraction of the club members are boys?
  (b) What percentage of the club members are girls?

6 Craig and Sophie share 40 chocolates.
They divide them in the ratio 1 : 4, with Sophie having the larger share.
How many chocolates does Sophie have?

AQA

7 Naheed is given £4. She spends £3.20 and saves the rest.
Express the amount she spends to the amount she saves as a ratio in its simplest form.

8 A pop concert is attended by 2100 people. The ratio of males to females is 2 : 3.
How many males attended the concert?

9 Dec shares a prize of £435 with Annabel in the ratio 3 : 2.
What is the difference in the amount of money they each receive?

10 The ratio of men to women playing golf one day is 5 : 3. There are 20 men playing.
How many women are playing?

11 A town has a population of 45 000 people. 1 in every 180 people are disabled.
How many disabled people are there in the town?

AQA

12 This is a list of ingredients to make 12 rock cakes.
You have plenty of margarine, sugar,
fruit and spice but only 500 g of flour.
What is the largest number of rock cakes
you can make?

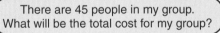

Rock cakes (makes 12)
240 g flour   150 g fruit
 75 g margarine $\frac{1}{4}$ teaspoon spice
125 g sugar

AQA

13 3 kg of pears cost £2.94.
How much will 2 kg of pears cost?

14 Two students are talking about their school outing.

My class went to Tower Bridge last week.
There are 30 people in my class.
The total cost was £345.

There are 45 people in my group.
What will be the total cost for my group?

15 A Munch Crunch bar weighs 21 g.
The table shows the nutrients that each bar contains.
  (a) What percentage of the bar is fat?
      Give your answer to an appropriate degree of accuracy.
  (b) What is the ratio of protein to carbohydrate?
      Give your answer in the form 1 : n.

Protein	1.9 g
Fat	4.7 g
Carbohydrate	13.3 g
Fibre	1.1 g

AQA

16 On a map the distance between two towns is 5 cm.
The actual distance between the towns is 1 kilometre.
What is the scale of the map in the form of 1 : n?

17 In a school, there are 750 pupils in total in years 9, 10 and 11.
The numbers of pupils in years 9, 10 and 11 are in the ratio 12 : 7 : 6.
How many pupils are there in each year?

AQA

Speed and Other Compound Measures

What you need to know

- **Speed** is a compound measure because it involves **two** other measures.

- **Speed** is a measurement of how fast something is travelling.
  It involves two other measures, **distance** and **time**.
  In situations where speed is not constant, **average speed** is used.

  $$\text{Speed} = \frac{\text{Distance}}{\text{Time}}$$

  $$\text{Average speed} = \frac{\text{Total distance travelled}}{\text{Total time taken}}$$

  The formula linking speed, distance and time can be rearranged and remembered as:
  $$S = D \div T$$
  $$D = S \times T$$
  $$T = D \div S$$

- You should be able to solve problems involving speed, distance and time.

  Eg 1 Wyn takes 3 hours to run 24 km. Calculate his speed in kilometres per hour.

  $$\text{Speed} = \frac{\text{Distance}}{\text{Time}} = \frac{24}{3} = 8 \text{ km/h}$$

  Eg 2 Norrie says, "If I drive at an average speed of 60 km/h it will take me $2\frac{1}{2}$ hours to complete my journey."
  What distance is his journey?

  $$\text{Distance} = \text{Speed} \times \text{Time} = 60 \times 2\frac{1}{2} = 150 \text{ km}$$

  Eg 3 Ellen cycles 5 km at an average speed of 12 km/h.
  How many minutes does she take?

  $$\text{Time} = \frac{\text{Distance}}{\text{Speed}} = \frac{5}{12} \text{ hours} = \frac{5}{12} \times 60 = 25 \text{ minutes}$$

  To change hours to minutes: **multiply by 60**

- **Density** is a compound measure which involves the measures **mass** and **volume**.

  Eg 4 A block of metal has mass 500 g and volume 400 cm³.

  $$\text{Density} = \frac{\text{Mass}}{\text{Volume}} = \frac{500}{400} = 1.25 \text{ g/cm}^3$$

  $$\text{Density} = \frac{\text{Mass}}{\text{Volume}}$$

- **Population density** is a measure of how populated an area is.

  Eg 5 The population of Cumbria is 489 700.
  The area of Cumbria is 6824 km².

  $$\text{Population density} = \frac{\text{Population}}{\text{Area}}$$

  $$\text{Population density} = \frac{\text{Population}}{\text{Area}} = \frac{489\ 700}{6824} = 71.8 \text{ people/km}^2.$$

Exercise 11

Do not use a calculator for questions 1 to 5.

1. Norma travels 128 km in 2 hours.
   Calculate her average speed in kilometres per hour.

2. Sean cycled 24 km at an average speed of 16 km/h.
   How long did he take to complete the journey?

3. Ahmed takes $2\frac{1}{2}$ hours to drive from New Milton to London. He averages 66 km/h.
   What distance does he drive?

4. Nigel runs 4 km at an average speed of 6 km/h.
   How many minutes does he take?

5 (a) Brian travels 225 miles by train. His journey takes $2\frac{1}{2}$ hours.
What is the average speed of the train?

(b) Val drives 225 miles at an average speed of 50 mph.
How long does her journey take?   AQA

6 Paul takes 15 minutes to run to school. His average running speed is 8 km/h.
How far did he have to run?

7 The distance between Heysham and the Isle of Man is 80 km.
A hovercraft travels at 50 km per hour. How long does the journey take?   AQA

8 A bus travels 12 miles in 45 minutes.
Calculate the average speed in miles per hour.   AQA

9 (a) An athlete runs 15 miles at an average speed of 6 miles per hour.
How long does he take to run the 15 miles?

(b) Another athlete runs 18 miles in $2\frac{1}{4}$ hours.
What is her average speed?   AQA

10 Kay walks 2.5 km in 50 minutes.
Calculate her average walking speed in kilometres per hour.

11 Sheila lives 6 kilometres from the beach.
She jogs from her home to the beach at an average speed of 10 km/h.
She gets to the beach at 1000.
Calculate the time when she left home.   AQA

12 The diagram shows the distances, in miles, between some junctions on a motorway.

West ◀───────25───── **12** ─────26───── **8** ─────27───── East▶

A coach is travelling west. At 1040 it passes junction 27 and at 1052 it passes junction 26.
(a) Calculate the average speed of the coach in miles per hour.

Between junctions 26 and 25 the coach travels at an average speed of 30 miles per hour.
(b) Calculate the time when the coach passes junction 25.

13 A train travels at an average speed of 80 miles per hour.
At 0940 the train is 65 miles from Glasgow.
The train is due to arrive in Glasgow at 1030.
Will it arrive on time? Show your working.

14 On Monday it took Helen 40 minutes to drive to work.
On Tuesday it took Helen 25 minutes to drive to work.
Her average speed on Monday was 18 miles per hour.
What was her average speed on Tuesday?

15 Henry completes a 200 m race in 25 seconds.
What is his average speed in kilometres per hour?   AQA

16 A jet-ski travels 0.9 kilometres in 1.5 minutes.
Calculate the average speed of the jet-ski in metres per second.

17 A copper statue has a mass of 1080 g and a volume of 120 cm³.
Work out the density of copper.

18 A silver medal has a mass of 200 g. The density of silver is 10.5 g/cm³.
What is the volume of the medal?

19 The population of Jamaica is 2.8 million people. The area of Jamaica is 10 800 km².
What is the population density of Jamaica?

Do not use a calculator for this exercise.

1 (a) (i) Write these numbers in order of size, smallest first: 16   10   6   100   61
(ii) What is the total when the numbers are added together?
(b) Work out. (i) $100 - 37$ (ii) 100×20 (iii) $100 \div 4$

2 (a) Work out $40 \times 50 \times 500$. Give your answer in words.
(b) What is the value of the 3 in the number 2439?

3 (a) Write the number   two thousand and thirty-six   in figures.
(b) Write (i) $\frac{1}{2}$ as a decimal,   (ii) $\frac{1}{4}$ as a percentage.
(c) Write the number 638 (i) to the nearest 10,   (ii) to the nearest 100.   AQA

4 Here is a set of numbers: 3,   5,   6,   9,   15,   21.
(a) Which two of these numbers have a product of 15?
(b) Which two of these numbers have a difference of 6 **and** a sum of 12?   AQA

5 (a) Work out. (i) $105 - 30$ (ii) 19×7 (iii) $2002 \div 7$
(b) Work out the square of 11.
(c) Work out. (i) $8 - 3 \times 2$ (ii) $(8 - 3) \times 2$

6 (a) A list of numbers is given.   4   5   6   12   24   36   45
(i) Which of these numbers is a factor of 18?
(ii) Which of these numbers is a multiple of 8?
(iii) Which of these numbers is a prime number?
(b) What are the common factors of 24 and 36?
(c) What is the square root of 36?

7 Orange juice is sold in cartons of two different sizes.

(a) How much is saved by buying a 500 ml carton instead of two 250 ml cartons?

(b) Reg buys four 500 ml cartons.
He pays with a £5 note.
How much change is he given?

8 An overnight train leaves Dundee at 2348 and arrives in London at 0735 the next day.
How long does the journey take? Give your answer in hours and minutes.

9 Kim states that the product of two consecutive whole numbers is an odd number.
By means of an example, show that Kim is **not** correct.   AQA

10 Alison worked for 2 hours 40 minutes and was paid £6 per hour.
How much did Alison earn?   AQA

11 (a) 4 litres of milk costs £2.96. How much is 1 litre of milk?
(b) Apples cost 98 pence per kilogram. What is the cost of 5 kilograms of apples?

12 To buy a car, Ricky has to pay 24 monthly payments of £198.
How much does he have to pay altogether to buy the car?

13 A group of 106 students travel to London by minibuses to watch an international hockey match.
Each minibus can carry 15 passengers.
What is the smallest number of minibuses needed?   AQA

14 Clive is making some four-digit numbers. Each number contains all the digits 3, 5, 9 and 6.
    (a) Write down the smallest number Clive can make.
    (b) Write down the largest **even** number Clive can make.   AQA

15 Samir is baking apple pies and gooseberry pies.
He uses 1.25 kg of flour in the apple pies. He uses 0.92 kg of flour in the gooseberry pies.
How much flour will he have left from a 3 kg bag of flour?   AQA

16 A ski-run measures 7.5 cm on a map. The map is drawn to a scale of 1 cm to 200 m.
What is the actual length of the ski-run in metres?

17 How much will it cost to hire a trailer for 5 days?

TRAILERS FOR HIRE
£3.50 per day
plus £12.50 insurance

18 A sports club is given £100 to spend on new footballs.
A new football costs £7.99.
What is the greatest number of footballs they can buy?

19 (a) Calculate the cost per litre of emulsion paint, correct to the nearest penny.
    (b) How much more does it cost to buy
10 litres of gloss paint than
10 litres of emulsion paint?

£12.95   GLOSS PAINT 5 litres   EMULSION PAINT 10 litres   £14.99

20 (a) Write 34.849, correct to one decimal place.
    (b) Work out $4.7 - 2.81$.   AQA

21 (a) The temperature at 6 am is $-3°C$. The temperature at 6 pm is $5°C$.
How many degrees warmer is it at 6 pm than at 6 am?
    (b) Work out $(-3) \times (-4)$.

22 One dollar is worth about 62 pence. Mr Jones buys a watch that costs 89 dollars.
Estimate how much the watch is worth in pounds.   AQA

23 A plank of wood is 225 cm in length. It is cut into two pieces.
One piece is 37 cm longer than the other. What is the length of the shorter piece of wood? AQA

24 (a) Work out (i) 10^5,   (ii) $10^2 - 2^5$,   (iii) $2^3 \times 3^2$,   (iv) $30^2 \div 10^3$.
    (b) Which is smaller, 5^4 or 4^5? Show **all** your working.
    (c) Work out $\sqrt{25} \times \sqrt{100}$.

25 (a) Write 0.7 as a fraction.
    (b) A turkey costs £3.60 per kg. What is the cost of a turkey which weighs 6.5 kg?
    (c) Work out. (i) 0.2×0.4 (ii) $24 \div 0.3$

26 (a) Write these fractions in order, smallest first: $\frac{1}{2}$ $\frac{2}{3}$ $\frac{3}{5}$ $\frac{5}{8}$ $\frac{3}{4}$
    (b) Write down a fraction that lies halfway between $\frac{1}{5}$ and $\frac{1}{4}$.
    (c) Work out (i) $\frac{1}{4} + \frac{2}{5}$,   (ii) $\frac{2}{3} - \frac{1}{2}$,   (iii) $\frac{4}{5} \times \frac{2}{3}$.
    (d) Work out $\frac{2}{5}$ of 12.

27 A crowd of 54 000 people watch a carnival.
    (a) 15% of the crowd are men. How many men watch the carnival?
    (b) Two-thirds of the crowd are children. How many children watch the carnival?

28 (a) Given that $59 \times 347 = 20\,473$, find the exact value of $\frac{20\,473}{590}$.
    (b) Use approximations to estimate the value of 49×302. Show all your working.

29 A packet contains 12 fibre-tipped pens. Henry and Alice share them in the ratio 1 : 3.
How many does Alice receive?   AQA

30 (a) Diesel costs £0.95 per litre in England. Calculate the cost of 45 litres of diesel.
 (b) In France, diesel is 20% cheaper than in England.
 Calculate the cost of 45 litres of diesel in France.

31 Colin buys two cups of tea and three cups of coffee. He pays £4.75 altogether.
 The price of a cup of tea is 89 pence. What is the price of a cup of coffee?   AQA

32 The cost of 6 medium eggs is 72 pence.
 (a) How much will 10 medium eggs cost?
 (b) Small eggs cost $\frac{7}{8}$ of the price of medium eggs. How much will 6 small eggs cost?
 (c) Large eggs cost 25% more than medium eggs. How much will 6 large eggs cost?   AQA

33 (a) A machine stamps 120 letters per minute.
  How long will it take to stamp 300 letters?
  Give your answer in minutes and seconds.
 (b) A machine sorts 2000 letters per hour at normal speed.
  At high speed it sorts 15% more letters per hour.
  How many letters per hour does it sort at high speed?   AQA

34 Four cabbages cost £2.88. How much will five cabbages cost?   AQA

35 (a) 7 *Arctic* ice creams cost a total of £6.65. How much will 12 *Arctic* ice creams cost?
 (b) A large box of ice cream cones contains 750 cones to the nearest 10.
  (i) What is the smallest possible number of cones in the box?
  (ii) What is the largest possible number of cones in the box?   AQA

36 (a) Conrad cycles 24 km in $1\frac{1}{2}$ hours. What is his cycling speed in kilometres per hour?
 (b) Cas cycles 24 km at 15 km/h. She sets off at 0930. At what time does she finish?

37 A quiz has 40 questions.
 (a) Lenny gets 34 questions right. What percentage of the questions did he get right?
 (b) 12 boys took part in the quiz. The ratio of boys to girls taking part in the quiz is 3 : 5.
  How many girls took part?

38 (a) Which of these numbers are prime?   25   29   33   37   48
 (b) Work out $2^3 \times 5^2$.
 (c) Evaluate $\frac{3}{4} - \frac{2}{5}$,   giving your answer as a fraction.
 (d) Given that $56 \times 473 = 26\,488$,   find the exact value of $\frac{264.88}{5.6}$.
 (e) Find an approximate value of $\frac{29.7 \times 5.98}{0.32}$.   Show your working.   AQA

39 A train travels from Basingstoke to London in 40 minutes. The distance is 50 miles.
 Find the average speed of the train in miles per hour.   AQA

40 Write as a single power of 3:   (a) $3^4 \times 3^3$ (b) $3^{10} \div 3^5$ (c) $\frac{3 \times 3^3}{3^2}$

41 Alan is $1\frac{3}{5}$ metres in height.   The height of a tree is $3\frac{1}{3}$ metres higher than Alan.
 Work out the height of the tree.

42 Trevor did a sponsored walk. The length of the walk, to the nearest kilometre, was 12 km.
 Trevor's time was 110 minutes to the nearest minute.
 (a) What was the longest length the walk could have been?
 (b) What was Trevor's shortest possible time?   AQA

43 A youth club organises a skiing holiday for 45 children. The ratio of boys to girls is 5 : 4.
 40% of the boys have skied before. How many boys have skied before?

44 (a) Write as a product of its prime factors: (i)   48,   (ii)   108.
 (b) Hence, find the least common multiple of 48 and 108.

Number
Calculator Paper

You may use a calculator for this exercise.

1 (a) Which of the numbers 8, −4, 0 or 5 is an odd number?
  (b) Write the number 3568 to the nearest 10.
  (c) What is the value of the 4 in the number 3.42?

2 (a) List these numbers in order, smallest first.

> 13   5   −7   0   −1

  (b) What is the difference between the largest number and the smallest number in your list?

3 Giles buys a newspaper for 55p and a computer magazine for £2.10.
  What change will he get from a £5 note?

4 Isaac buys 180 grams of sweets from the Pic 'n' Mix selection.
  The price of the sweets is 65p per 100 g.
  How much does he have to pay?

5 In a long jump event Hanniah jumped the following distances.
> 5.15 m   4.95 m   5.20 m   5.02 m   5.10 m
  (a) Write down the shortest distance Hanniah jumped.
  (b) Write these distances in order, shortest first.

6 Some of the rail services from Poole to Waterloo are shown.

Poole	0544	0602	—	0640	—	0740	0825	0846
Bournemouth	0558	0616	—	0654	0715	0754	0839	0900
Southampton	0634	0655	0714	0738	0754	0838	0908	0938
Eastleigh	0646	—	—	0750	—	0852	—	0951
Waterloo	0804	0810	0844	0901	0908	1005	1018	1112

  (a) Sid arrives at Bournemouth station at 0830.
    What is the time of the next train to Eastleigh?
  (b) Paul catches the 0654 from Bournemouth to Southampton.
    How many minutes does the journey take?

7 Eric earns £563 per month. He has a tax allowance of £4895 per year.
  (a) How much is Eric's taxable income per year?

  He pays tax at the rate of 10p in the pound.
  (b) How much tax does Eric pay per year?   AQA

8 A quiz consists of ten questions. Beth, John and Sue take part. These are their results.

	Beth	John	Sue
Number of answers correct	4	6	5
Number of answers incorrect	4	3	1
Number of questions not attempted	2	1	4

  A correct answer scores 3 points. An incorrect answer scores −2 points.
  A question not attempted scores 0 points.
  Who scores the most points? Show your working.   AQA

9 (a) Write $\frac{7}{9}$ as a decimal.   Give your answer correct to two decimal places.
  (b) Write 33%, 0.3, $\frac{8}{25}$ and $\frac{1}{3}$ in order of size, smallest first.

10 Bruce buys two packets of baby wipes on special offer.
Calculate the actual cost of
each baby wipe.

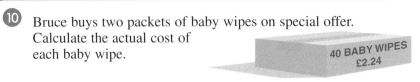

40 BABY WIPES
£2.24

Special Offer
BUY ONE
GET ONE FREE

11 Cheri is paid a basic rate of £6.40 per hour for a 35-hour week.
Overtime is paid at $1\frac{1}{2}$ times the basic rate.
Last week she worked 41 hours. Calculate her pay for last week.

12 On a musical keyboard there are 5 black keys for every 7 white keys.
The keyboard has 28 white keys. How many black keys does it have?

13 Work out.   (a) $6.25 \times 13 - 6.25 \times 3$ (b) $\frac{2}{5}$ of 18   (c) $3.5^2 - 2.5^2$

14 A packet of washing powder costs £4.18 and weighs 1.5 kg.
The packet has enough powder for 22 washes.
(a)   What is the cost of powder for one wash?
(b)   How much powder is needed for one wash?
Give your answer in grams correct to one decimal place.   AQA

15 (a)   James earns £9650 a year. He gets a pay rise of 6%.
(i)   How much **more** does James earn after the pay rise?
(ii)   What is James' new pay per year?
(b)   Frances is paid £11 400 a year. She takes home $\frac{4}{5}$ of her pay.
(i)   How much does Frances take home each year?
(ii)   Frances is paid monthly. How much does she take home each month?   AQA

16 Jacob is 3.7 kg heavier than Isaac. The sum of their weights is 44.5 kg. How heavy is Jacob?

17 In America a camera cost $110.
In England an identical camera costs £65.
The exchange rate is £1 = $1.62
In which country is the camera cheaper and by how much?
You must show all your working.

$110   £65

AQA

18 (a)   Write $\frac{13}{20}$ as a decimal.
(b)   In a spelling test Lara scores 13 out of 20. What is Lara's score as a percentage?

19 A shop sells 4000 items in a week. 5% are returned.
$\frac{1}{4}$ of the returned items are faulty. How many items are faulty?   AQA

20 A car takes $2\frac{1}{2}$ hours to travel 150 km.
Calculate the average speed of the car in kilometres per hour.

21 (a)   Use your calculator to find $2.4^3 + \sqrt{180}$.   Give all the figures on your calculator.
(b)   Write your answer to one significant figure.

22 To make squash, orange juice and water is mixed in the ratio of 1 : 6.
How much orange juice is needed to make 3.5 litres of squash?

23 Mrs Joy's electricity meter was read on 1st March and 1st June.
On 1st March the reading was $\boxed{3\,2\,4\,5\,7}$ On 1st June the reading was $\boxed{3\,2\,9\,3\,1}$
(a)   How many units of electricity have been used?

Her electricity bill for this period includes a fixed charge of £11.58 and the cost of the units
used at 9.36 pence per unit.
(b)   Calculate the total cost of electricity for this period.

24 Kelly states that $a^2 + b^2$ is always an even number when a and b are prime numbers.
By means of an example, show that Kelly is **not** correct.   AQA

SR

25 The diagram shows the weights and prices of two packets of gravy granules.
This week both packets are on special offer.
The smaller packet has one third off the normal price.
The larger packet has 30% off the normal price.
Which packet is better value this week? Show your working.

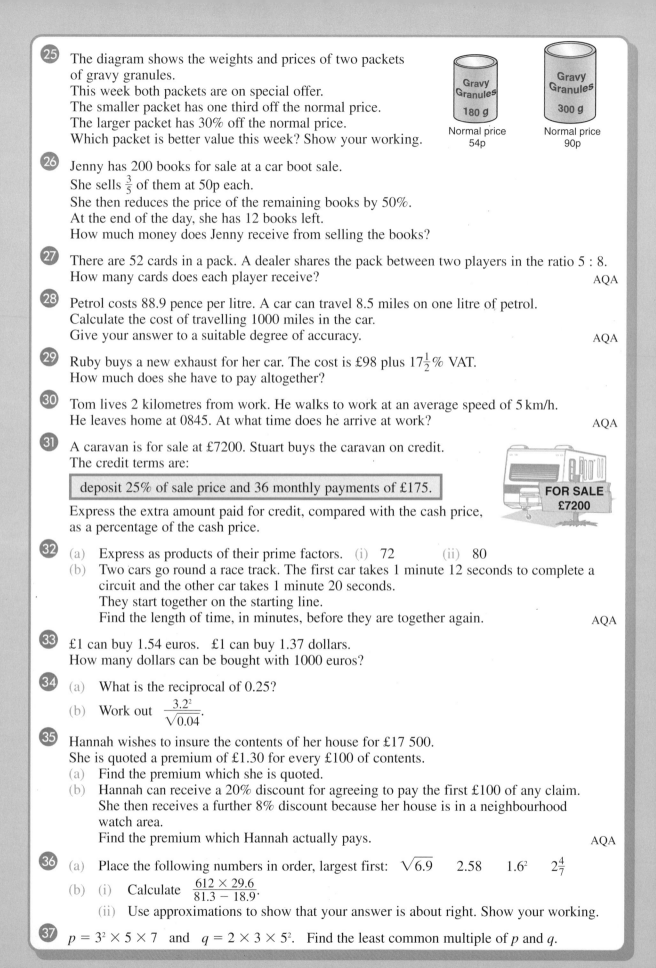

Gravy Granules 180 g
Normal price 54p

Gravy Granules 300 g
Normal price 90p

26 Jenny has 200 books for sale at a car boot sale.
She sells $\frac{3}{5}$ of them at 50p each.
She then reduces the price of the remaining books by 50%.
At the end of the day, she has 12 books left.
How much money does Jenny receive from selling the books?

27 There are 52 cards in a pack. A dealer shares the pack between two players in the ratio 5 : 8.
How many cards does each player receive?   AQA

28 Petrol costs 88.9 pence per litre. A car can travel 8.5 miles on one litre of petrol.
Calculate the cost of travelling 1000 miles in the car.
Give your answer to a suitable degree of accuracy.   AQA

29 Ruby buys a new exhaust for her car. The cost is £98 plus $17\frac{1}{2}$% VAT.
How much does she have to pay altogether?

30 Tom lives 2 kilometres from work. He walks to work at an average speed of 5 km/h.
He leaves home at 0845. At what time does he arrive at work?   AQA

31 A caravan is for sale at £7200. Stuart buys the caravan on credit.
The credit terms are:

> deposit 25% of sale price and 36 monthly payments of £175.

Express the extra amount paid for credit, compared with the cash price,
as a percentage of the cash price.

FOR SALE
£7200

32 (a) Express as products of their prime factors. (i) 72   (ii) 80
(b) Two cars go round a race track. The first car takes 1 minute 12 seconds to complete a circuit and the other car takes 1 minute 20 seconds.
They start together on the starting line.
Find the length of time, in minutes, before they are together again.   AQA

33 £1 can buy 1.54 euros.   £1 can buy 1.37 dollars.
How many dollars can be bought with 1000 euros?

34 (a) What is the reciprocal of 0.25?

(b) Work out $\dfrac{3.2^2}{\sqrt{0.04}}$.

35 Hannah wishes to insure the contents of her house for £17 500.
She is quoted a premium of £1.30 for every £100 of contents.
(a) Find the premium which she is quoted.
(b) Hannah can receive a 20% discount for agreeing to pay the first £100 of any claim.
She then receives a further 8% discount because her house is in a neighbourhood watch area.
Find the premium which Hannah actually pays.   AQA

36 (a) Place the following numbers in order, largest first: $\sqrt{6.9}$ 2.58 1.6^2 $2\frac{4}{7}$

(b) (i) Calculate $\dfrac{612 \times 29.6}{81.3 - 18.9}$.

(ii) Use approximations to show that your answer is about right. Show your working.

37 $p = 3^2 \times 5 \times 7$ and $q = 2 \times 3 \times 5^2$. Find the least common multiple of p and q.

Introduction to Algebra

What you need to know

- You should be able to write **algebraic expressions**.

  Eg 1   An expression for the cost of 6 pens at n pence each is $6n$ pence.

  Eg 2   An expression for 2 pence more than n pence is $n + 2$ pence.

- Be able to **simplify expressions** by collecting **like terms** together.

  Eg 3   (a) $2d + 3d = 5d$ (b) $3x + 2 - x + 4 = 2x + 6$ (c) $x + 2x + x^2 = 3x + x^2$

- Be able to **multiply expressions** together.

  Eg 4   (a) $2a \times a = 2a^2$ (b) $y \times y \times y = y^3$ (c) $3m \times 2n = 6mn$

- Recall and use these properties of powers:
  Powers of the same base are **added** when terms are **multiplied**.
  Powers of the same base are **subtracted** when terms are **divided**.

  $$a^m \times a^n = a^{m+n}$$
  $$a^m \div a^n = a^{m-n}$$

  Eg 5   (a) $x^3 \times x^2 = x^5$ (b) $a^5 \div a^2 = a^3$

- Be able to **multiply out brackets**.

  Eg 6   (a) $2(x + 5) = 2x + 10$ (b) $x(x - 5) = x^2 - 5x$

  (c) $(x + 2)(x + 5) = x^2 + 5x + 2x + 10 = x^2 + 7x + 10$

- Be able to **factorise expressions**.

  Eg 7   (a) $3x - 6 = 3(x - 2)$ (b) $m^2 + 5m = m(m + 5)$

Exercise 12

1   A calculator costs £9.
Write an expression for the cost of k calculators.

2   Godfrey is 5 years older than Mary.
Write an expression for Godfrey's age when Mary is t years old.

3   Simplify the following.

(a) $3x + 2x - x$ (b) $5x + 3y - 2x + 4y$ (c) $3 \times a \times 4$ AQA

4   A cup of coffee costs x pence and a cup of tea costs y pence.
Write an expression for the cost of 3 cups of coffee and 2 cups of tea.

5   Simplify.

(a) $m + 2m + 3m$ (b) $2m + 2 - m$ (c) $m \times m \times m$

6   (a)   A pen costs x pence. How much will 4 pens cost?
(b)   Simplify $3m + 2 + m - 3$.   AQA

7   Write an expression, in terms of x,
for the sum of the angles in this shape.

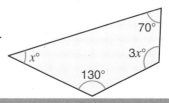

8 A man leads a string of n donkeys.

(a) How many heads are there?
(b) How many legs are there?

AQA

9 A muffin costs $d + 3$ pence.
Write an expression for the cost of 5 muffins.

10 Simplify the expressions.

(a) $2p + p + 5p$ (b) $t \times t \times t$ (c) $5x + y - 2x + 3y$ AQA

11 Which of these algebraic expressions are equivalent?

$2y$	y^2	$2(y + 1)$	$y \times y$	$y + y$
$2y + 2$	$2y + y$	$2y^2$	$3y$	$2y + 1$

AQA

12 (a) Simplify $2x + y - x + y$.

(b) Multiply out (i) $2(x + 3)$, (ii) $x(x - 1)$.

13 Simplify (a) $5a \times 2a$, (b) $3g \times 2h$, (c) $6k \div 3$, (d) $3m \div m$.

14 Grace is x years old.
Harry is 3 years older than Grace.

(a) Write down an expression for Harry's age.
(b) Jack is twice as old as Harry.
Write down an expression for Jack's age. AQA

15 (a) Simplify $5p + 2q - q + 2p$.

(b) Multiply out (i) $4(r - 3)$, (ii) $s(s^2 + 6)$. AQA

16 (a) Ken works x hours a week for £y per hour.
Write an expression for the amount he earns each week.

(b) Sue works 5 hours less than Ken each week and earns £y per hour.
Write an expression for the amount Sue earns each week.

17 (a) Simplify $8a + ab - a + 2b + 3ab$.

(b) Expand and simplify $5(x + 3) - x$. AQA

18 Multiply out and simplify $3(4x - 1) + 2x - 6$. AQA

19 (a) Simplify $9p + 4q + 6p - 7q$.

(b) Multiply out and simplify $8(x + 3) + 5(2x + 4)$. AQA

20 Simplify. (a) $y^3 \times y^2$ (b) $x^6 \div x^3$ (c) $\dfrac{z^4 \times z}{z^3}$

21 Factorise. (a) $3x - 6$ (b) $x^2 - 2x$ AQA

22 (a) Simplify (i) $5 - 3(2n - 1)$, (ii) $3(2x + 3) - 2(5 + x)$.

(b) Multiply out and simplify $(p - 3)(p + 4)$.

23 Simplify. (a) $a \times a \times a \times a$ (b) $b \times b^3$ (c) $\dfrac{c^5}{c^2}$ (d) $\dfrac{d^3 \times d^5}{d^4}$

24 (a) Factorise $4x + 6$.

(b) Expand (i) $3(2y - 3)$, (ii) $x(x^2 - 2x)$, (iii) $a(a + b)$.

(c) Simplify $2x^2 - x(1 + x)$.

(d) Multiply out and simplify $(m - 2)(m - 3)$.

What you need to know

- The solution of an equation is the value of the unknown letter that fits the equation.

- You should be able to solve simple equations by **inspection**.

  Eg 1 (a) $a + 2 = 5$ (b) $m - 3 = 7$ (c) $2x = 10$ (d) $\frac{t}{4} = 3$

  $a = 3$ $m = 10$ $x = 5$ $t = 12$

- Be able to solve simple problems by **working backwards**.

  Eg 2 I think of a number, multiply it by 3 and add 4. The answer is 19.

  x → | multiply by 3 | → | add 4 | → Answer 19

  5 ← | divide by 3 | ← 15 ← | subtract 4 | ← 19

  The number I thought of is 5.

- Be able to use the **balance method** to solve equations.

  Eg 3 Solve these equations.

  (a) $d - 13 = -5$ (b) $-4a = 20$ (c) $5 - 4n = -1$

  $d = -5 + 13$ $a = \frac{20}{-4}$ $-4n = -6$

  $d = 8$ $a = -5$ $n = 1.5$

Exercise 13

1 What number should be put in the box to make each of these statements correct?

(a) $\boxed{} - 6 = 9$ (b) $2 + \boxed{} = 11$ (c) $4 \times \boxed{} = 20$ (d) $\dfrac{\boxed{}}{5} = 3$

2 Solve these equations.

(a) $7 + x = 12$ (b) $5 - x = 3$ (c) $\frac{x}{2} = 7$ (d) $3x = 21$

3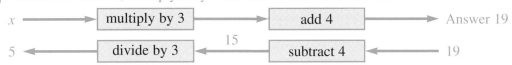

Teacher: Think of a number, double it and add 5.

Zeenat

The number I thought of was 25.

John: My answer was 19.

(a) What answer did Zeenat get?
(b) What was the number John thought of?

AQA

4 (a) I think of a number, add 3, and then multiply by 2.
The answer is 16. What is my number?

(b) I think of a number, double it and then subtract 3.
The answer is 5. What is my number?

5 Solve these equations.

(a) $3x - 7 = 23$ (b) $4 + 3x = 19$ (c) $5x - 9 = 11$ (d) $5 - 7x = 47$

6 Solve these equations.

(a) $3x + 5 = 2$ (b) $4x = 2$ (c) $4x + 1 = 23$ (d) $5x + 1 = -3$

Further Equations

What you need to know

- To solve an equation you need to find the numerical value of the letter, by ending up with **one letter** on one side of the equation and a **number** on the other side of the equation.

- You should be able to solve equations with unknowns on both sides of the equals sign.

  Eg 1 Solve $3x + 1 = x + 7$.
  $$3x = x + 6$$
  $$2x = 6$$
  $$x = 3$$

- Be able to solve equations which include brackets.

  Eg 2 Solve $2(x - 3) = 4$.
  $$2x - 6 = 4$$
  $$2x = 10$$
  $$x = 5$$

- You should be able to solve equations using a **trial and improvement** method.
  The value of the unknown letter is improved until the required degree of accuracy is obtained.

  Eg 3 Use a trial and improvement method to find a solution to the equation $x^3 + x = 40$,
  correct to one decimal place.

x	$x^3 + x$	Comment
3	$27 + 3 = 30$	Too small
4	$64 + 4 = 68$	Too big
3.5	$42.8\ldots + 3.5 = 46.3\ldots$	Too big
3.3	$35.9\ldots + 3.3 = 39.2\ldots$	Too small
3.35	$37.5\ldots + 3.35 = 40.9\ldots$	Too big

  > For accuracy to 1 d.p.
  > check the second decimal place.
  > The solution lies between
  > 3.3 and 3.35.

  $x = 3.3$,   correct to 1 d.p.

- You should be able to write, or form, equations using the information given in a problem.

Exercise 14

1 Solve these equations.   (a) $3 + x = 7$ (b) $3 - x = 4$ (c) $3x = 15$ (d) $\frac{x}{3} = 7$

2 Matt thinks of a number.   He multiplies it by 4 and then takes away 5.   The answer is 39.
What was the number?   AQA

3 In the table below, the letters w, x, y and z represent different numbers.
The total of each row is given at the side of the table.

w	w	w	w	24
w	w	x	x	28
w	w	x	y	25
w	x	y	z	23

Find the values of w, x, y and z.   AQA

4 Solve the equations   (a) $3x - 7 = x + 15$,   (b) $5(x - 2) = 20$.

5 Solve these equations.
(a) $7x + 4 = 60$ (b) $3x - 7 = -4$ (c) $2(x + 3) = -2$ (d) $3x - 4 = 1 + x$

6 Solve the equations   (a) $2x + 3 = 15$,   (b) $3(x - 1) = 6$,   (c) $x + 2 = 5 - x$.
AQA

7 Solve these equations.
(a) $2x + 5 = 2$ (b) $2(x - 1) = 3$ (c) $5 - 2x = 3x + 2$ (d) $2(3 + x) = 9$

8 Solve the equation $7y - 1 = 3 - y$.
AQA

9 The lengths of these rods are given, in centimetres, in terms of n.

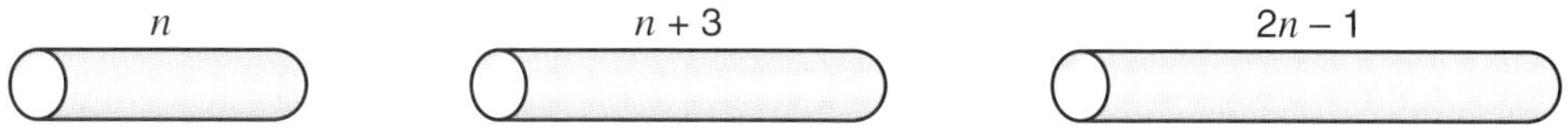

n $n + 3$ $2n - 1$

(a) Write an expression, in terms of n, for the total length of the rods.
(b) The total length of the rods is $30\,cm$.
By forming an equation, find the value of n.

10 Mandy buys a small box of chocolates and a large box of chocolates.
The diagram shows the number of chocolates in each box.
Altogether there are 47 chocolates.
By forming an equation, find the number
of chocolates in the larger box.

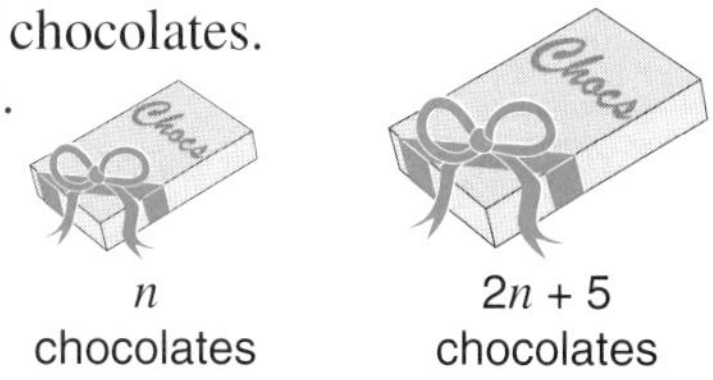

n $2n + 5$
chocolates   chocolates

11 A café sells cakes and buns.
(a) Write down an expression for the cost, in pence, of y cakes at 40p each
and 3 buns at 60p each.
(b) The total cost of y cakes and 3 buns is £4.60.
Find the number of cakes sold.
AQA

12 Solve the equation $4(3 - x) = 20$.

13 Solve these equations.   (a) $3x - 5 = 16$ (b) $7x + 1 = 2x + 4$

14 Solve the equation $5(x - 3) = 2x$.

15 The total of each row is given at the side of the table.
Find the values of x and A.

$4x + 1$	$2(x + 5)$	20
$2x$	4	A

AQA

16 Solve these equations.
(a) $\frac{x}{3} = -7$ (b) $5x - 3 = 7$ (c) $2(y + 5) = 3$ (d) $z + 7 = 3 - 4z$

17 Solve the equations   (a) $\frac{x - 7}{5} = 2$,   (b) $5x + 6 = 24 - 10x$.
AQA

18 Solve the equations.
(a) $\frac{20}{x} = 4$ (b) $\frac{y}{3} + 5 = 9$ (c) $4(z - 1) = 2(z + 3)$
AQA

19 Dario is using trial and improvement to find a solution to the equation $x + \frac{1}{x} = 5$.
The table shows his first trial.

x	$x + \frac{1}{x}$	**Comment**
4	4.25	Too low

Continue the table to find a solution to the equation.
Give your answer to 1 decimal place.
AQA

Formulae ●●●●●●●●●●●●●●●

What you need to know

- An **expression** is just an answer using letters and numbers.
  A **formula** is an algebraic rule. It always has an equals sign.

- You should be able to **write simple formulae**.

  Eg 1 A packet of crisps weighs 25 grams.
  Write a formula for the total weight, W grams, of n packets of crisps.
  $$W = 25n$$

  Eg 2 Start with t, add 5 and then multiply by 3.
  The result is p.
  Write a formula for p in terms of t.
  $$p = 3(t + 5)$$

- Be able to **substitute** values into expressions and formulae.

  Eg 3 (a) Find the value of
  $4x - y$ when
  $x = 5$ and $y = 7$.
  $$\begin{aligned} 4x - y &= 4 \times 5 - 7 \\ &= 20 - 7 \\ &= 13 \end{aligned}$$

  (b) $A = pq - r$
  Find the value of
  A when $p = 2$,
  $q = -2$ and $r = 3$.
  $$\begin{aligned} A &= pq - r \\ A &= 2 \times (-2) - 3 \\ A &= -4 - 3 \\ A &= -7 \end{aligned}$$

  (c) $M = 2n^2$
  Find the value of
  M when $n = 3$.
  $$\begin{aligned} M &= 2n^2 \\ M &= 2 \times 3^2 \\ M &= 2 \times 9 \\ M &= 18 \end{aligned}$$

- Be able to **rearrange** a simple formula to make another letter (variable) the subject.

  Eg 4 $y = 2x + 5$. Make x the subject of the formula.
  $$y = 2x + 5$$
  Take 5 from both sides. $y - 5 = 2x$
  Divide both sides by 2. $\dfrac{y - 5}{2} = x$.   So, $x = \dfrac{y - 5}{2}$

Exercise 15   Do not use a calculator for questions 1 to 14.

1 What is the value of $a - 3b$ when $a = 10$ and $b = 2$?

2 What is the value of $2x + y$ when $x = -3$ and $y = 5$?

3 (a) Work out $4p - 1$ when $p = 10$.
   (b) Work out $8a + 4b$ when (i) $a = 2$ and $b = 3$,
                    (ii) $a = 2$ and $b = -3$.   AQA

4 Given that $m = -3$ and $n = 5$, find the value of
   (a) $m + n$,   (b) $m - n$,   (c) $n - m$,   (d) mn.

5 $H = ab - c$.   Find the value of H when $a = 2$, $b = -5$ and $c = 3$.

6 If $x = 5$ and $y = -7$, find the value of (a) $4x + 3y$,   (b) $\dfrac{x - y}{4}$.   AQA

7 A boat is hired. The cost, in £, is given by:

$$\boxed{\text{Cost} = 6 \times \text{Number of hours} + 5}$$

(a) Calculate the cost of hiring the boat for 2 hours.
(b) The boat was hired at a cost of £29. For how many hours was it hired? AQA

8 $L = 5(p + q)$. Find the value of L when $p = 2$ and $q = -4$.

9 $A = b - cd$. Find the value of A when $b = -3$, $c = 2$ and $d = 4$.

10 If $p = 4$ and $q = -5$, find the value of (a) $3pq$, (b) $p^2 + 2q$.

11 (a) Find the value of $5p + 2q$ when $p = 4$ and $q = -7$.
(b) Find the value of $u^2 - v^2$ when $u = 5$ and $v = 3$. AQA

12 What is the value of $10y^2$ when $y = 3$?

13 What is the value of $3x^3$ when $x = 2$?

14 $T = ab^2$. Find the value of T when $a = 4$ and $b = -5$.

15 The charges on a light railway are worked out by this formula.

$$\boxed{\text{30p per mile plus 25p}}$$

(a) Abdul travels 7 miles on the railway. How much is he charged?
(b) Belle is charged £3.85. How far does she travel? AQA

16 Using $p = 18.8$, $q = 37.2$, $r = 0.4$, work out: (a) $p + \dfrac{q}{r}$ (b) $\dfrac{p + q}{r}$ AQA

17 This rule is used to change miles into kilometres.

$$\boxed{\text{Multiply the number of miles by 8 and then divide by 5}}$$

(a) Use the rule to change 25 miles into kilometres.
(b) Using K for the number of kilometres and M for the number of miles write a formula for K in terms of M.
(c) Use your formula to find the value of M when $K = 60$.

18 The formula $F = \dfrac{9}{5}C + 32$ is used to change temperatures in degrees Centigrade (C) to temperatures in degrees Fahrenheit (F).
A thermometer reads a temperature of 15°C.
What is the equivalent temperature in degrees Fahrenheit? AQA

19 Given that $m = \dfrac{1}{2}$, $p = \dfrac{3}{4}$, $t = -2$, calculate (a) $mp + t$, (b) $\dfrac{(m + p)}{t}$. AQA

20 A formula is given as $c = 3t - 5$. Rearrange the formula to give t in terms of c.

21 (a) Make c the subject of the formula $P = 2a + 2b + 2c$.
(b) Find the value of c when $P = 46$, $a = 7.7$ and $b = 10.8$. AQA

22 A formula for calculating distance is $d = \dfrac{(u + v)t}{2}$.
Find the value of d when $u = 9.4$, $v = 6.3$ and $t = 8$. AQA

23 $m = 3(n - 17)$. Find the value of n when $m = -9$.

24 Make r the subject of the formula $p = \dfrac{5r}{s}$.

25 You are given the formula $v = u + at$.
(a) Find v when $u = 17$, $a = -8$ and $t = 3$.
(b) Rearrange the formula to give a in terms of v, u and t.

Sequences ● ● ● ● ● ● ● ● ● ● ● ●

What you need to know

- A **sequence** is a list of numbers made according to some rule.
  The numbers in a sequence are called **terms**.

- You should be able to draw and continue number sequences represented by patterns of shapes.

  Eg 1 This pattern represents the sequence: 3, 5, 7, …

- Be able to continue a sequence by following a given rule.

  Eg 2 The sequence 2, 7, 22, … is made using the rule:

  > Multiply the last number by 3, then add 1.

  The next term in the sequence = $(22 \times 3) + 1 = 66 + 1 = 67$

- Be able to find a rule, and then use it, to continue a sequence.

  > **To continue a sequence:**
  > 1. Work out the rule to get from one term to the next.
  > 2. Apply the same rule to find further terms in the sequence.

  Eg 3 Describe the rule used to make the following sequences.
  Then use the rule to find the next term of each sequence.
  - (a) 5, 8, 11, 14, …   Rule: add 3 to last term.   Next term: 17.
  - (b) 2, 4, 8, 16, …   Rule: multiply last term by 2. Next term: 32.
  - (c) 1, 1, 2, 3, 5, 8, … Rule: add the last two terms.   Next term: 13.

- Special sequences   **Square numbers:**   1, 4, 9, 16, 25, …
  Triangular numbers: 1, 3, 6, 10, 15, …

- A number sequence which increases (or decreases) by the same amount from one term to the next is called a **linear sequence**.
  The sequence: 2, 8, 14, 20, 26, … has a **common difference** of 6.

- You should be able to find an expression for the nth term of a linear sequence.

  Eg 4 Find the nth term of the sequence: 3, 5, 7, 9, …
  The sequence is linear, common difference = 2.
  To find the nth term add one to the multiples of 2.
  So, the nth term is $2n + 1$.

Exercise 16

1 Write down the next two terms in each of these linear sequences.
  - (a) 1, 5, 9, 13, 17, …
  - (b) 50, 46, 42, 38, 34, …

2 What is the next number in each of these sequences?
  - (a) 1, 2, 5, 10, …
  - (b) 1, 3, 9, 27, …
  - (c) 1, $\frac{1}{2}$, $\frac{1}{4}$, $\frac{1}{8}$, …

3 The first six terms of a sequence are shown. 1, 4, 5, 9, 14, 23, …
  Write down the next two terms.

4 Sasha makes a sequence of patterns with sticks. Here are his first three patterns.

 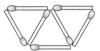

Pattern 1　　**Pattern 2**　　**Pattern 3**

(a) Draw Pattern 4.

(b) Copy and complete the table.

Pattern number	1	2	3	4
Number of sticks	3	5		

(c) How many sticks will Sasha use for Pattern 5?

(d) There are 33 sticks in Pattern 16.
How many sticks are in Pattern 17?

AQA

5 Look at this sequence of numbers. 2, 5, 8, 11, …

(a) What is the next number in the sequence?

(b) Is 30 a number in this sequence? Give a reason for your answer.

6 The rule for a sequence is:

> Add the last two numbers and divide by 2.

Write down the next three terms when the sequence begins: 3, 7, …

7 A sequence begins: 5, 15, 45, 135, …

(a) Write down the rule, in words, used to get from one term to the next in the sequence.

(b) Use your rule to find the next term in the sequence.

8 A sequence begins: 1, −2, … The next number in the sequence is found by using the rule:

> ADD THE PREVIOUS TWO NUMBERS AND MULTIPLY BY TWO

Use the rule to find the next **two** numbers in the sequence.　　AQA

9 Ahmed writes down the first four numbers of a sequence: 10, 8, 4, −2, …

(a) What is the next number in this sequence?

(b) Explain how you found your answer.

AQA

10 A sequence begins: 1, 6, 10, 8, … The rule to continue the sequence is:
double the difference between the last two numbers.
Ravi says if you continue the sequence it will end in 0. Is he correct? Explain your answer.

11 The first three patterns in a sequence are shown.

Pattern 1　　　　**Pattern 2**　　　　**Pattern 3**

(a) How many squares are in pattern 20? Explain how you found your answer.

(b) Write an expression for the number of squares in the nth pattern.

12 A sequence is given by 5, 12, 19, 26, 33, …

(a) What is the next term in this sequence? Explain how you got your answer.

(b) Write down the nth term for the sequence.　　AQA

13 Find the nth term of the following sequences.

(a) 5, 7, 9, 11, …

(b) 1, 5, 9, 13, …

14 (a) Write down the first **three** terms of the sequence whose nth term is given by $n^2 + 4$.

(b) Will the number 106 be in this sequence? Explain your answer.　　AQA

What you need to know

- **Coordinates** (involving positive and negative numbers) are used to describe the position of a point on a graph.

  Eg 1 The coordinates of A are $(4, 1)$.
  The coordinates of B are $(-3, 2)$.

- The x axis is the line $y = 0$. The y axis is the line $x = 0$.

- The x axis crosses the y axis at the **origin**.

- The graph of a linear function is a straight line.

- You should be able to draw the graph of a straight line.

  Eg 2 Draw the graphs of the following lines.

  (a) $y = 2$ (b) $x = 3$ (c) $y = \frac{1}{2}x + 1$

The graph is a **horizontal** line. All points on the line have y coordinate 2.	The graph is a **vertical** line. All points on the line have x coordinate 3.	Find values for x and y.

x	0	2	4
y	1	2	3

  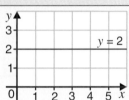 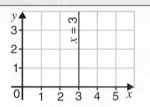 

- Be able to draw the graph of a straight line by finding the points where the line crosses the x axis and the y axis.

  Eg 3 Draw the graph of the line $x + 2y = 4$.

  At the point where a graph crosses:
  the x axis, $y = 0$,
  the y axis, $x = 0$.

  When $y = 0$, $x + 0 = 4$, $x = 4$. Plot $(4, 0)$.
  When $x = 0$, $0 + 2y = 4$, $y = 2$. Plot $(0, 2)$.
  A straight line drawn through the points $(0, 2)$ and $(4, 0)$ is the graph of $x + 2y = 4$.

  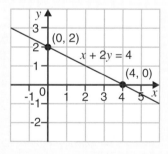

- The equation of the graph of a straight line is of the form $y = mx + c$, where m is the gradient and c is the y-intercept.

  The **gradient** of a line can be found by drawing a right-angled triangle.

  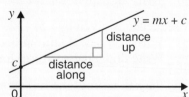

  $$\text{Gradient} = \frac{\text{distance up}}{\text{distance along}}$$

  Gradients can be positive, zero or negative.

You should be able to:

- interpret the graph of a linear function,
- use the graphs of linear functions to solve equations.

Exercise **17**

1 The diagram shows the line segment *RS*.

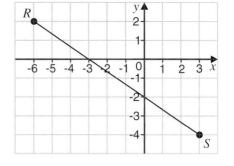

(a) Write down the coordinates of points *R* and *S*.
(b) The straight line joining *R* and *S* crosses
the *x* axis at *T* and the *y* axis at *U*.
Write down the coordinates of *T* and U.

2 Draw and label *x* and *y* axes from −5 to 4.
(a) On your diagram plot *A* (4, 3) and *B* (−5, −3).
(b) *C* (*p*, −1) is on the line segment *AB*.
What is the value of *p*?

3 (a) On the same diagram draw the lines $y = 2$ and $x = 5$.
(b) Write down the coordinates of the point where the lines cross.

4 The diagram shows a sketch of the line $2y = 6 − x$.

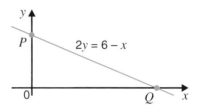

(a) Find the coordinates of the points *P* and *Q*.
(b) The line $2y = 6 − x$ goes through *R* (−5, *m*).
What is the value of *m*?

5 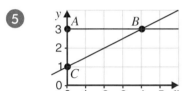 Points *A*, *B* and *C* are shown on the grid.
(a) Write down the equation of the line *AB*.
(b) Use the grid to work out the equation of the line *CB*.

AQA

6 (a) Copy and complete the table of values for $y = 1 − 2x$.

x	−3	0	3
y		1	

(b) Draw the line $y = 1 − 2x$ for values of *x* from −3 to 3.
(c) Use your graph to find the value of *y* when $x = −1.5$.

7 On the same diagram, draw and label the lines: $y = x + 1$ and $y = 1 − x$.   AQA

8 (a) On the same axes, draw the graphs of $y = −2$, $y = x$ and $x + y = 5$.
(b) Which of these lines has a negative gradient?

9 On separate diagrams draw the graphs of each of these equations for values of *x* from −2 to 2.
(a) $y = 2x$ (b) $y − x = 2$ (c) $y + x = 2$ (d) $2y = x$

10 (a) Copy and complete the table of values for $2y = 3x − 6$.

x	−2	0	4
y		−3	

(b) Draw the graph of $2y = 3x − 6$ for values of *x* from −2 to 4.
(c) Use your graph to find the value of *x* when $y = 1.5$.

11 (a) Draw the graph of $5y − 2x = 10$ for values of *x* from −5 to 5.
(b) Use your graph to find the value of *y* when $x = −2$.

12 (a) On the same diagram, draw and label the lines $y = x − 1$ and $x + y = 5$
for values of *x* from 0 to 5.
(b) Write down the coordinates of the point where the lines cross.

Coordinates and Graphs

What you need to know

- A graph used to change from one quantity into an equivalent quantity is called a **conversion graph**.

  Eg 1 Use 15 kilograms = 33 pounds (lb) to draw a conversion graph for kilograms and pounds.

  Use your graph to find (a) 5 kilograms in pounds, (b) 20 pounds in kilograms.

  The straight line drawn through the points (0, 0) and (33, 15) is the conversion graph for kilograms and pounds.

  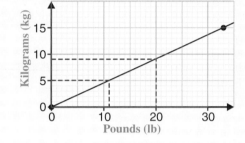

  Reading from the graph:
  (a) 5 kg = 11 lb
  (b) 20 lb = 9 kg

- **Distance-time graphs** are used to illustrate journeys.

  On a distance-time graph:
  Speed can be calculated from the gradient of a line.
  The faster the speed the steeper the gradient.
  Zero gradient (horizontal line) means zero speed.

  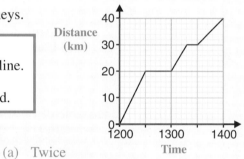

  Eg 2 The graph shows a car journey.
  (a) How many times does the car stop?   (a) Twice
  (b) (i) Between what times does the car travel fastest? Explain your answer.
       (ii) What is the speed of the car during this part of the journey?
  (b) (i) 1200 to 1230. Steepest gradient.
       (ii) Speed = $\dfrac{\text{Distance}}{\text{Time}} = \dfrac{20\,\text{km}}{\frac{1}{2}\,\text{hour}} = 40\,\text{km/h}$

- You should be able to draw and interpret graphs arising from real-life situations.

Exercise 18

1 This graph can be used to convert miles to kilometres.

(a) Scott lives $3\frac{1}{2}$ miles from the Post Office.
How many kilometres is this?

(b) Jade goes for a training run of 7 kilometres.
How many miles is this?

(c) Jade is training to run a half marathon, which is a distance of 13 miles.
Use the graph to calculate this distance in kilometres.

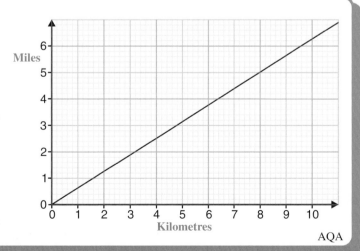

AQA

2 The graph shows the temperature of the water in a tank as it is being heated.

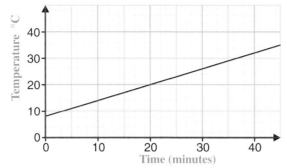

(a) What was the temperature of the water before it was heated?

(b) How long did it take for the water to reach 26°C?

(c) Estimate the number of minutes it will take for the temperature of the water to rise from 32°C to 50°C.

3 (a) Given that 7.4 square metres = 80 square feet, draw a conversion graph for square metres to square feet.

(b) Use your graph to change
   (i) 5 square metres into square feet,
   (ii) 32 square feet into square metres.

4 The table shows the largest quantity of salt, w grams, which can be dissolved in a beaker of water at temperature t°C.

t°C	10	20	25	30	40	50	60
w grams	54	58	60	62	66	70	74

(a) Draw a graph to illustrate this information.

(b) Use your graph to find
   (i) the lowest temperature at which 63 g of salt will dissolve in the water,
   (ii) the largest amount of salt that will dissolve in the water at 44°C.

(c) (i) The equation of the graph is of the form $w = at + b$.
       Use your graph to estimate the values of the constants a and b.
   (ii) Use the equation to calculate the largest amount of salt which will dissolve in the water at 95°C.

AQA

5 A salesman is paid a basic amount each month plus commission on sales.
The graph shows how the monthly pay of the salesman depends on his sales.

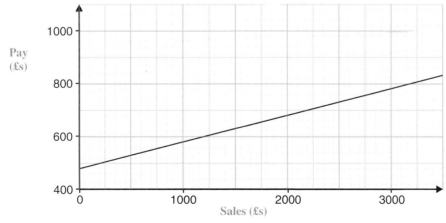

(a) How much is the salesman's monthly basic pay?

(b) How much commission is the salesman paid on £1000 of sales?

(c) Calculate the monthly pay of the salesman when his sales are £6400.

6 The distance by boat from Poole Quay to Wareham is 12 miles.
The distance-time graph shows a boat trip from Poole Quay to Wareham and back.

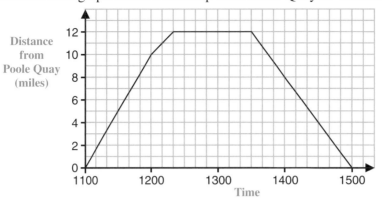

(a) Describe what happened to the speed of the boat at 1200 hours.
(b) How long did the boat stay in Wareham?
(c) What was the average speed of the boat on the return journey from Wareham to
Poole Quay?

AQA

7 Ken drives from his home to the city centre.
The graph represents his journey.
(a) How long did Ken take to reach the city centre?
(b) How far from the city centre does Ken live?
(c) What is his average speed for the journey in
kilometres per hour?

8 The graph shows a train journey from *A* to *D*, stopping at *B* and *C*.

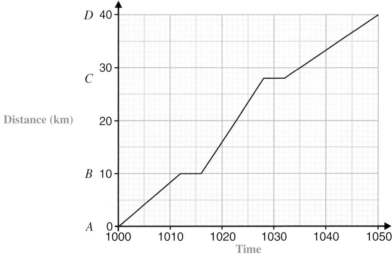

(a) What is the time when the train leaves *B*?
(b) How far is it from *A* to *C*?
(c) During which part of the journey did the train travel the fastest? Explain your answer.
(d) Calculate the speed of the train from *A* to *B* in kilometres per hour.

AQA

9 Water is poured at a constant rate into each
of the two containers shown.
Sketch graphs to show the height of water
in the containers as they are being filled.

Inequalities

What you need to know

- **Inequalities** can be described using words or numbers and symbols.

Sign	Meaning
<	is less than
≤	is less than or equal to

Sign	Meaning
>	is greater than
≥	is greater than or equal to

- Inequalities can be shown on a **number line**.

Eg 1 This diagram shows the inequality: $-2 < x \leq 3$

The circle is: **filled** if the inequality is **included** (i.e. ≤ or ≥),
not filled if the inequality is **not included** (i.e. < or >).

- **Solving inequalities** means finding the values of x which make the inequality true.

Eg 2 Solve these inequalities.
  (a) $3x < 6$
     $x < 2$
  (b) $x + 3 \geq 5$
     $x \geq 2$
  (c) $7a \geq a + 9$
     $6a \geq 9$
     $a \geq 1.5$

Eg 3 Find the integer values of n for which $-1 \leq 2n + 3 < 7$.
$$-1 \leq 2n + 3 < 7$$
$$-4 \leq 2n < 4$$
$$-2 \leq n < 2$$

Integer values which satisfy the inequality $-1 \leq 2n + 3 < 7$ are: $-2, -1, 0, 1$

Exercise 19

1 Solve these inequalities.
  (a) $5x > 15$
  (b) $x + 3 \geq 1$
  (c) $x - 5 \leq 1$
  (d) $3 + 2x > 7$

2 Draw number lines to show each of these inequalities.
  (a) $x \geq -2$
  (b) $\frac{x}{3} < -1$
  (c) $-1 < x \leq 3$
  (d) $x \leq -1$ **and** $x > 3$

3 (a) Solve the inequality $3x - 2 \leq 7$.
  (b) Show the solution to (a) on a number line.

4 Solve these inequalities.
  (a) $2x \leq 6 - x$
  (b) $3x > x + 7$
  (c) $5x < 2x - 4$

5 List the values of n, where n is an integer such that:
  (a) $-2 \leq 2n < 6$
  (b) $-3 < n - 3 \leq -1$
  (c) $-5 \leq 2n - 3 < 1$

6 List the integer values of x such that $6 \leq 5x < 20$.

7 (a) Solve the inequality $2x + 3 \geq 1$.
  (b) Write down the inequality shown by the following diagram.

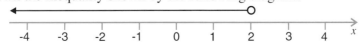

  (c) Write down all the integers that satisfy both inequalities shown in parts (a) and (b). AQA

Quadratic Graphs

What you need to know

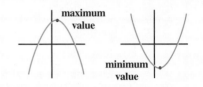

- The graph of a **quadratic function** is a **smooth curve**.
- The general equation for a **quadratic function** is
  $y = ax^2 + bx + c$, where a cannot be zero.
  The graph of a quadratic function is symmetrical
  and has a **maximum** or **minimum** value.

You should be able to:
- **substitute** values into given functions to generate points,
- plot graphs of **quadratic functions**,
- use graphs of quadratic functions to solve equations.

Eg 1
(a) Draw the graph of $y = x^2 - 2x - 5$ for values of x from -2 to 4.
(b) Use your graph to solve the equation $x^2 - 2x - 5 = 0$.

(a)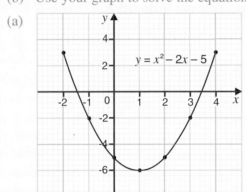

To draw a quadratic graph:
Make a table of values connecting x and y.
Plot the points.
Join the points with a smooth curve.

x	-2	-1	0	1	2	3	4
y	3	-2	-5	-6	-5	-2	3

To solve the equation, read the values of x where the graph of $y = x^2 - 2x - 5$ crosses the x axis $(y = 0)$.

(b) $x = -1.4$ and 3.4, correct to one decimal place.

Exercise 20

1 (a) Copy and complete this table of values for $y = x^2 - 2$.

x	-3	-2	-1	0	1	2	3
y		2		-2	-1		7

  (b) Draw the graph of $y = x^2 - 2$ for values of x from -3 to 3.
  (c) Write down the values of x at the points where the line $y = 3$ crosses your graph.
  (d) Write down the values of x where $y = x^2 - 2$ crosses the x axis.

2 (a) Draw the graph of $y = x^2 + x - 2$ for values of x from -2 to 3.
  (b) State the minimum value of y.
  (c) Use your graph to solve the equation $x^2 + x - 2 = 0$.

3 A sky diver jumps from a plane. The table shows the distance he falls, d metres, in t seconds.

t (seconds)	0	0.5	1.5	2.5	3.5
d (metres)	0	1	11	31	61

  (a) Plot these points and join them with a smooth curve.
  (b) Use your graph to find how many seconds he takes to fall 50 m.
  (c) Use your graph to estimate how far he has fallen after 4 seconds.

AQA

Algebra
Non-calculator Paper

Do not use a calculator for this exercise.

1 What are the coordinates of *A*?

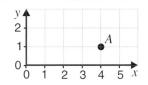

2 A sequence begins   2,   4,   6,   …
To continue the sequence use the rule:

> Add 2 to the last term.

(a)   Write down the next term in the sequence.
(b)   Explain why the number 99 is not a term in this sequence.

3 In each part, find the output when the input is 12.

(a)

(b)

4 Use this rule to find the number of points a football team has scored.

> Points scored = 3 × Number of wins + Number of draws

A team wins 7 games and draws 5. How many points have they scored?

5 Regular pentagons are used to form patterns, as shown.

Pattern 1   **Pattern 2**   **Pattern 3**

(a)   Draw Pattern 4.
(b)   Copy and complete the table.

Pattern number	1	2	3	4
Number of sides	5	8	11	

(c)   How many sides has Pattern 5?
(d)   Pattern 10 has 32 sides. How many sides has Pattern 11?

6 Find the value of $3a + 2b$ when $a = 5$ and $b = 3$.

7 (a)   On graph paper, plot the points $A(-3, -2)$ and $B(1, 4)$.
(b)   What are the coordinates of the midpoint of the line segment *AB*?

8 A jam doughnut costs *t* pence.
(a)   Write an expression for the cost of 5 jam doughnuts.

A cream doughnut costs 5 pence more than a jam doughnut.
(b)   Write an expression for the cost of a cream doughnut.

9 What number should be put in the box to make each of these statements correct?

(a) $\square - 3 = 7$ (b) $\square + 5 = 9$ (c) $3 \times \square = 6$ (d) $\dfrac{\square}{3} = 5$

10 (a)   (i)   What is the next term in this sequence?   2,   9,   16,   23,   …
(ii)   Will the 50th term in the sequence be an odd number or an even number?
Give a reason for your answer.
(b)   Another sequence begins   1,   5,   9,   13,   17,   …
Describe in words the rule for continuing the sequence.

11 $2n$ represents any even number.
Which of the statements describes the number   (a) n,   (b) $2n + 1$?
always even   always odd   could be even or odd

12 Nick thinks of a number.   He doubles it and then subtracts 3.   The answer is 17.
What is his number?

13 A large envelope costs x pence and a small envelope costs y pence.
Write an expression for the cost of 3 large envelopes and 5 small envelopes.

14 Which of these algebraic expressions are equivalent?

| $2a - a$ | $3a$ | $2(a - 1)$ | $2a + a$ |
| $2a + 1$ | $2a - 2$ | $a + a - 1$ | 2 |

15 Simplify   (a) $7x - 5x + 3x$,   (b) $a - 3b + 2a - b$,   (c) $3 \times m \times m$.

16 (a)   Find the value of $\dfrac{3(m + 9)}{n}$ when $m = -5$ and $n = 24$.

(b)   Find the value of $3p + q$ when $p = -2$ and $q = 5$.

17 (a)   Simplify $3n - n + 5$.
(b)   Work out the value of $2x + y^3$ when $x = -3$ and $y = 2$.   AQA

18 Here is a flow diagram.   Input → Subtract 5 → Multiply by -3 → Output

(a)   What is the output when the input is 3?
(b)   What is the input when the output is -21?   AQA

19 Solve these equations.   (a) $2x + 6 = 14$ (b) $3g - 5 = 4$ (c) $10y = 5$ (d) $6 + 2y = 4$

20 The line $y = -3$ crosses the line $y = x - 2$ at the point P.
What are the coordinates of P?   AQA

21 The travel graph shows the journey of a train from London to Manchester.

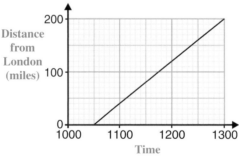

Calculate the average speed of the train in miles per hour.   AQA

22 Find the value of $3x + y^3$ when $x = -1$ and $y = -2$.

23 Hannah is x years old.
(a)   Her sister Louisa is 3 years younger than Hannah.
Write an expression, in terms of x, for Louisa's age.
(b)   Their mother is four times as old as Hannah.
Write an expression, in terms of x, for their mother's age.
(c)   The total of their ages is 45 years.
By forming an equation in x, find their ages.

24 (a)   Factorise   (i) $3a - 6$,   (ii) $k^2 - 2k$.
(b)   Multiply out   (i) $5(x + 3)$,   (ii) $m(m - 4)$.
(c)   Solve   (i) $3 - 4x = x + 8$,   (ii) $3(2x + 1) = 6$.

25 Given that $s = 2t^3$, find the value of t when $s = 250$.

26 $y = \frac{4}{5}(9 - x)$. Find the value of x when $y = 6$.

27 (a) Solve the equation $\frac{3x + 5}{2} = 7$.

(b) Solve the inequality $4x - 8 < 12$ and show the solution on a number line.

28 (a) On the same diagram draw the graphs $2y = x + 4$ and $y = \frac{1}{2}x + 1$.
(b) What do you notice about the two lines you have drawn?

29 A pencil costs x pence. A crayon costs $x + 3$ pence.
(a) Write an expression in terms of x for the cost of 5 crayons.

One pencil and 5 crayons cost 87 pence.
(b) By forming an equation in x, find the cost of the pencil.   AQA

30 Matches are arranged to form a sequence of diagrams as shown.

        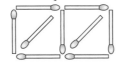   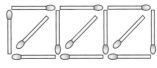

   Diagram 1   **Diagram 2**   **Diagram 3**

Write an expression, in terms of n, for the number of matches needed to form the nth diagram.
                                                                         AQA

31 (a) Solve the inequality $3x < 6 - x$.
(b) List all the values of n, where n is an integer, such that $-3 < 2x + 1 \leq 3$.

32 Match these equations to their graphs.

A $y = x$
B $y + x = 1$
C $y = x^2$
D $y = x^3$

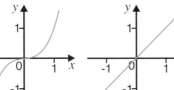   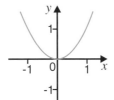   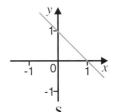

   P   **Q**   **R**   **S**

33 (a) Solve the equation $x - 4 = 3(x + 1)$.
(b) Simplify $y^3 \times y^5$.   AQA

34 (a) A sequence begins: $-2, \ 1, \ 4, \ 7, \ 10, \ \ldots$
Write, in terms of n, the nth term of the sequence.
(b) Make x the subject of the formula $y = 2x - 5$.

35 (a) Copy and complete the table of values for $y = x^2 - 2x + 1$.

x	-1	0	1	2	3
y		1	0		4

(b) Draw the graph of $y = x^2 - 2x + 1$ for values of x from -1 to 3.
(c) Use your graph to solve the equations
(i) $x^2 - 2x + 1 = 0$,   (ii) $x^2 - 2x + 1 = 2$.

36 Water is poured into a container at a constant rate.
Copy the axes given and sketch the graph of the
depth of the water against time as the container is filled.

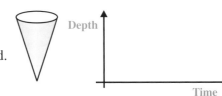

37 (a) Multiply out and simplify $(p - 2)(p + 2)$.

(b) Simplify $\dfrac{q^3 \times q^5}{q^2}$.

Algebra
Calculator Paper

You may use a calculator for this exercise.

1 (a) A sequence begins: 1, 2, 4, 7, 11, 16, ...
  (i) What is the next term in the sequence?
  (ii) Describe the rule you used to find the next term.
(b) Another sequence uses this rule:

> Add 3 to the last term.

What term comes before 15 in this sequence?

2 (a) Simplify. (i) $y + 2y + 3y$ (ii) $3m + 4 + m - 3$
(b) Work out the missing values in these calculations.
  (i) $10 \rightarrow \boxed{-3} \rightarrow \boxed{\times 2} \rightarrow$ ....   (ii) .... $\rightarrow \boxed{-3} \rightarrow \boxed{\times 2} \rightarrow 10$

3 (a) A ball costs x pence. How much will 3 balls cost?
(b) A skipping rope costs 30 pence more than a ball. How much does a skipping rope cost?

4 Donna and Todd work at SupaMotors selling cars.
Their weekly pay is calculated using this formula.

> Pay = £50 for each car sold plus £75.

(a) Donna sells 5 cars in one week. Work out her pay for this week.
(b) Last week Todd's pay was £625. How many cars did he sell last week?   AQA

5 (a) Write down the next **two** terms in each of these sequences:
  (i) 1, 3, 5, 7, 9, ...   (ii) 1, 2, 4, 8, 16, ...
(b) Use the sequences in part (a) to find the next two terms in this sequence:
  2, 5, 9, 15, 25, ...   AQA

6 In this 'algebraic' magic square, every row, column and diagonal should add up and
simplify to $15a + 12b + 6c$.

$8a + 5b + 5c$	$a + 6b - 2c$	$6a + b + 3c$
	$5a + 4b + 2c$	$7a + 8b + 4c$
$4a + 7b + c$		$2a + 3b - c$

(a) Copy and complete the magic square.
(b) Calculate the value of $15a + 12b + 6c$ if $a = 1$, $b = 2$ and $c = 3$.   AQA

7 This conversion graph can be used to change euros to dollars.

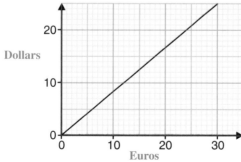

(a) Use the graph to find (i) 30 euros in dollars, (ii) 15 dollars in euros.
(b) Explain how you can use the graph to change 100 dollars into euros.

8 A sequence begins: 1, 2, 6, 16, … This is the rule continuing the sequence.

> ADD THE PREVIOUS TWO NUMBERS TOGETHER AND THEN MULTIPLY BY TWO

Deepak says the next term in the sequence is 22. Is he correct? **Explain your answer.**   AQA

9 (a) Copy and complete the table of values for the equation $y = x - 2$.

x	-1	1	3
y		-1	

(b) Draw the graph of $y = x - 2$ for values of x from -1 to 3.
(c) What are the coordinates of the points where the graph crosses the x axis and the y axis?

10 Solve these equations. (a) $g - 5 = 3$ (b) $4 + a = 9$ (c) $7x = 42$ (d) $5x + 4 = 19$

11 Umbrellas cost £4 each.
(a) Write a formula for the cost, C, in pounds, of u umbrellas.
(b) Find the value of u when $C = 28$.

12 (a) Find the value of $3m - 5$ when $m = 4$.
(b) $T = 3m - 5$. Find the value of m when $T = 4$.
(c) $P = 5y^2$. Find the value of P when $y = 3$.

13 Jaspel says to his friends:

> "Think of a number, add 5, then divide by 2. Tell me your answer."

(a) Huw thinks of the number 7. What is his answer?
(b) Gary says his answer is 11. What number did he start with?   AQA

14 I think of a number. If I double my number and add 1, my answer is 35.
(a) Write down an equation to describe this.
(b) What number am I thinking of?   AQA

15 The graph shows the journey of a cyclist from Halton to Kendal.
The distance from Halton to Kendal is 30 miles.

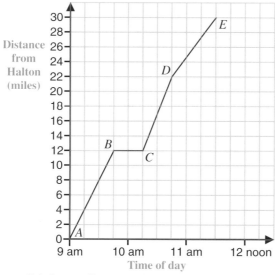

(a) For how long did the cyclist stop during the journey?
(b) What was the average speed for the part of the journey from A to B?
(c) On which section of the journey was the cyclist travelling at his fastest speed?
    Explain clearly how you got your answer.
(d) The cyclist stayed in Kendal for 2 hours.
    He then returned to Halton, without stopping, at an average speed of 12 miles per hour.
    Calculate the time he arrived back in Halton.   AQA

16 Here is a rule for working out a sequence of numbers.

| Choose a starting number S | → | Multiply by 3 | → | Subtract 4 | → | Write down the final number F. |

Write down an **equation** connecting the final number, F, and the starting number, S.　　AQA

17 (a) Draw the line $y = 2x + 1$ for values of x from -1 to 2.
(b) The line $y = 2x + 1$ crosses the line $x = -5$ at P. Give the coordinates of P.

18 (a) Solve. (i) $\frac{x}{2} = 10$ (ii) $7 - 6x = 8x$ (b) Multiply out. $3(t - 4)$

19 The nth term of a sequence is $3n - 1$.
(a) Write down the first and second terms of the sequence.
(b) Which term of the sequence is equal to 32?
(c) Explain why 85 is not a term in this sequence.　　AQA

20 (a) Solve the equations (i) $4(a - 2) = 6$, (ii) $5t + 3 = -1 + t$.
(b) The sum of the numbers x, $x - 3$ and $x + 7$ is 25.
By forming an equation in x, find the value of x.

21 (a) This rule is used to produce a sequence of numbers.

MULTIPLY THE LAST NUMBER BY 3 AND SUBTRACT 1

The second number in the sequence is 20. What is the first number?
(b) Another sequence begins 2, 5, 8, 11, …
(i) One number in the sequence is x.
Write, in terms of x, the next number in the sequence.
(ii) Write, in terms of n, the nth term of the sequence.　　AQA

22 Use trial and improvement to find the solution to the equation $x^3 - 3x = 9$.

x	$x^3 - 3x$	Comment
2	2	Too low
3	18	Too high

Copy and complete the table. Give your answer to one decimal place.　　AQA

23 A glass of milk costs x pence. A milk shake costs 45 pence more than a glass of milk.
(a) Write an expression for the cost of a milk shake.
(b) Lou has to pay £4.55 for 3 milk shakes and a glass of milk.
By forming an equation, find the price of a glass of milk.

24 (a) Copy and complete the table of values for $y = x^2 - 5$.

x	-3	-2	-1	0	1	2	3
y	4		-4	-5			4

(b) Draw the graph of $y = x^2 - 5$ for values of x from -3 to 3.
(c) Use your graph to solve the equation $x^2 - 5 = 0$.

25 A solution to the equation $x^3 + 2x = 90$ lies between 4 and 5.
Use trial and improvement to find this solution. Show all your trials.
Give your answer correct to one decimal place.

26 Expand and simplify. (a) $3(2x - 1) + 2(4x - 5)$ (b) $(x + 2)(x + 4)$　　AQA

27 (a) Solve the equation $3 - x = 4(x + 1)$.
(b) Multiply out and simplify $2(5x - 3) - 3(x - 1)$.
(c) Simplify (i) $m^8 \div m^2$, (ii) $n^2 \times n^3$.

28 Make x the subject of the formula $3x + y = 7$.

29 (a) List all the values of n, where n is an integer, such that $-2 \leqslant n - 3 < 1$.
(b) Factorise $xy - y^2$.　　AQA

What you need to know

- You should be able to use a **protractor** to measure and draw angles accurately.

  Eg 1 Measure the size of this angle.

  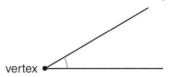

  vertex •

  The angle measures 30°.

  | To measure an angle, the protractor is placed so that its centre point is on the corner (vertex) of the angle, with the base along one of the arms of the angle, as shown. 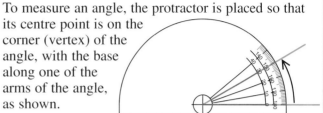 |

- Types and names of angles.

Acute angle	**Right angle**	**Obtuse angle**	**Reflex angle**
$0° < a < 90°$	$a = 90°$	$90° < a < 180°$	$180° < a < 360°$

- Angle properties.

Angles at a point	**Complementary angles**	**Supplementary angles**	**Vertically opposite angles**
		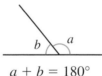	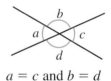
$a + b + c = 360°$	$x + y = 90°$	$a + b = 180°$	$a = c$ and $b = d$

- A straight line joining two points is called a **line segment**.

- Lines which meet at right angles are **perpendicular** to each other.

- Lines which never meet and are always the same distance apart are **parallel**.

- When two parallel lines are crossed by a **transversal** the following pairs of angles are formed.

Corresponding angles	**Alternate angles**	**Allied angles**	
			Arrowheads are used to show that lines are **parallel**.
$a = c$	$b = c$	$b + d = 180°$	

- You should be able to use angle properties to solve problems involving lines and angles.

  Eg 2 Work out the size of the angles marked with letters.
  Give a reason for each answer.

  $a + 64° = 180°$ (supplementary angles)
      $a = 180° - 64° = 116°$

     $b = 64°$ (vertically opposite angles)
     $c = 64°$ (corresponding angles)

  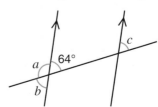

1 Look at the diagram.

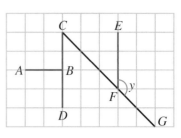

(a) Which lines are parallel to each other?
(b) Which lines are perpendicular to each other?
(c) (i) Measure angle *y*.
    (ii) Which of these words describes angle *y*?

      acute angle   **obtuse angle**   **reflex angle**

2 This shape contains a right angle, acute angles, obtuse angles and a reflex angle.

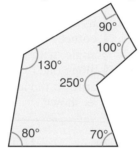

Write down the size of
(a) one of the acute angles,
(b) one of the obtuse angles,
(c) the reflex angle.

AQA

3 Without measuring, find the size of the lettered angles.
Give a reason for each of your answers.

(a)
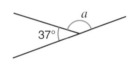

(b)
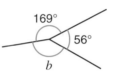

(c)
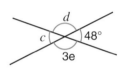

4
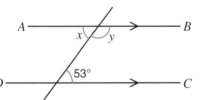

AB is parallel to *DC*.

(a) Work out the size of angle *x*.
    Give a reason for your answer.
(b) Work out the size of angle *y*.
    Give a reason for your answer.

5 In the diagram, the lines *PQ* and *RS* are parallel.

(a) What is the size of angle *x*?
    Give a reason for your answer.
(b) Find the size of angle *y*.

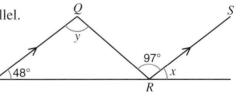

AQA

6
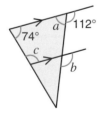

Work out the size of the angles marked with letters.
Give a reason for each answer.

7 Find the size of the angles marked with letters.

(a)
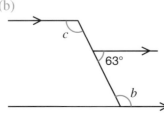

(b)
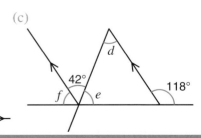

(c)

54

What you need to know

- Triangles can be: **acute-angled** (all angles less than 90°),
  obtuse-angled (one angle greater than 90°),
  right-angled (one angle equal to 90°).

- The sum of the angles in a triangle is 180°.
  $a + b + c = 180°$

- The exterior angle is equal to the sum of the two opposite interior angles. $a + b = d$

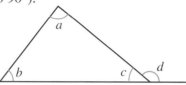

- Types of triangle:

| **Scalene** | **Isosceles** | **Equilateral** |

Sides have different lengths.
Angles are all different.

Two equal sides.
Two equal angles.

Three equal sides.
Three equal angles, 60°.

> A **sketch** is used when an accurate drawing is not required.
> Dashes across lines show sides that are equal in length.
> Equal angles are marked using arcs.

- You should be able to use properties of triangles to solve problems.

  Eg 1 Find the size of the angles marked a and b.
  $a = 86° + 51°$ (ext. $\angle$ of a $\triangle$)
  $a = 137°$
  $b + 137° = 180°$ (supp. $\angle$'s)
  $b = 43°$

  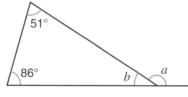

- Perimeter of a triangle is the sum of its three sides.

- Area of a triangle $= \dfrac{\text{base} \times \text{perpendicular height}}{2}$
  $$A = \tfrac{1}{2} \times b \times h$$

  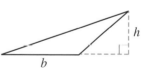
  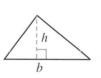

  Eg 2 Calculate the area of this triangle.
  $A = \tfrac{1}{2} \times b \times h$
  $ = \tfrac{1}{2} \times 9 \times 6\,\text{cm}^2$
  $ = 27\,\text{cm}^2$

  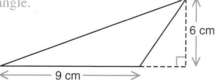

- You should be able to draw triangles accurately, using ruler, compasses and protractor.

Exercise 22

1 Without measuring, work out the size of the angles marked with letters.

(a)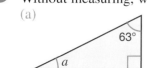

(b)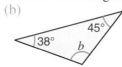

(c)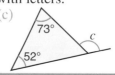

2 In the diagram, *ABX* is a straight line.
Work out the size of angle *ACB*.

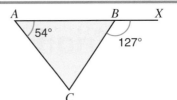

AQA

3

The diagram shows triangle *PQR*, with *PQ = PR*.
Work out the value of *x*.
Give a reason for your answer.

AQA

4 The diagram shows an isosceles triangle with two sides extended.
(a) Work out the size of angle *x*.
(b) Work out the size of angle *y*.

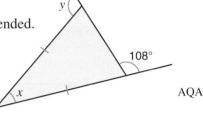

AQA

5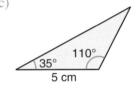

In the diagram *AB = BC = CA = CD*.
Work out the size of angle *CDA*.
Explain how you found your answer.

AQA

6 Make accurate drawings of these triangles using the information given.
(a) (b) (c)

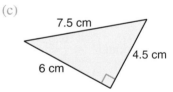

7 Find the areas of these triangles.
(a) (b) (c)

8 (a) Construct accurately a triangle with sides of 8 cm, 6 cm and 5 cm.
(b) By measuring the base and height, calculate the area of the triangle.

9 The diagram shows triangle *PQR*.
Calculate the area of triangle *PQR*.

AQA

10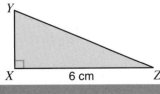

Triangle *XYZ* has an area of 12 cm².
XZ = 6 cm.
Calculate *YX*.

Symmetry and Congruence

What you need to know

- A two-dimensional shape has **line symmetry** if the line divides the shape so that one side fits exactly over the other.

- A two-dimensional shape has **rotational symmetry** if it fits into a copy of its outline as it is rotated through 360°.

- A shape is only described as having rotational symmetry if the order of rotational symmetry is 2 or more.

- The number of times a shape fits into its outline in a single turn is the **order of rotational symmetry**.

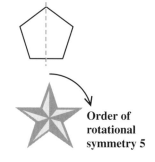

Order of rotational symmetry 5

Eg 1 For each of these shapes   (a)   draw and state the number of lines of symmetry,
(b)   state the order of rotational symmetry.

(i)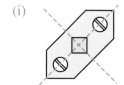

(ii)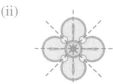

(iii)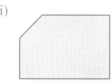

Two lines of symmetry.
Rotational symmetry of order 2.

4 lines of symmetry.
Order of rotational symmetry 4.

No lines of symmetry.
Order of rotational symmetry 1.
The shape is **not** described as having rotational symmetry.

- A **plane of symmetry** slices through a three-dimensional object so that one half is the mirror image of the other half.

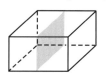

- Three-dimensional objects can have **axes of symmetry**.

Eg 2 Sketch a cuboid and show its axes of symmetry.

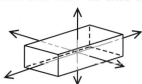

A cuboid has three axes of symmetry. The order of rotational symmetry about each axis is 2.

- When two shapes are the same shape and size they are said to be **congruent**.

- There are four ways to show that a pair of triangles are congruent.

SSS 3 corresponding sides.	**ASA** 2 angles and a corresponding side.
SAS 2 sides and the included angle.	**RHS** Right angle, hypotenuse and one other side.

Eg 3 Which of these triangles are congruent to each other? Give a reason for your answer.

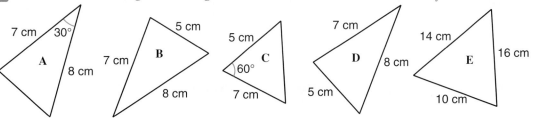

B and D.   Reason: 3 corresponding sides (SSS)

1 Half of a shape is drawn on squared paper, as shown.
AB is a line of symmetry for the complete shape.
Copy the diagram and complete the shape.

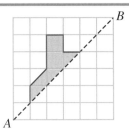

2 Consider the letters of the word **O R A N G E**

Which letters have   (a)   line symmetry only,
(b)   rotational symmetry only,
(c)   line symmetry and rotational symmetry?

3 For each of these shapes state   (i)   the number of lines of symmetry,
(ii)   the order of rotational symmetry.

(a)   (b)   (c)   (d)

    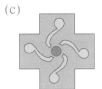   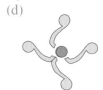

4

The diagram shows part of a shape.
Complete the shape so that it has
rotational symmetry of order 2.

AQA

5 The diagram shows a square-based pyramid.
(a)   How many planes of symmetry has the pyramid?
(b)   How many axes of symmetry has the pyramid?

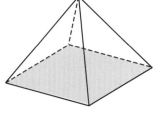

6

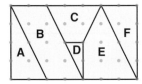

The diagram shows a rectangle which has been cut into 6 pieces.
Which two pieces are congruent to each other?

7 (a)   Complete this shape so that it has both line symmetry
and rotational symmetry.

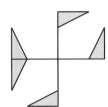

(b)   Which of these triangles are congruent to each other?
Give a reason for your answer.

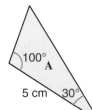   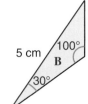   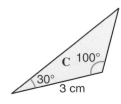   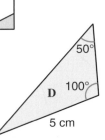

AQA

Quadrilaterals

What you need to know

- A **quadrilateral** is a shape made by four straight lines.
- The sum of the angles in a quadrilateral is 360°.
- The **perimeter** of a quadrilateral is the sum of the lengths of its four sides.

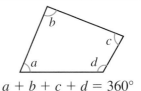

$$a + b + c + d = 360°$$

- Facts about these special quadrilaterals:

rectangle square parallelogram rhombus trapezium isosceles trapezium kite

Quadrilateral	Sides	Angles	Diagonals	Line symmetry	Order of rotational symmetry	Area formula
Rectangle	Opposite sides equal and parallel	All 90°	Bisect each other	2	2	$A = \text{length} \times \text{breadth}$ $A = lb$
Square	4 equal sides, opposite sides parallel	All 90°	Bisect each other at 90°	4	4	$A = (\text{length})^2$ $A = l^2$
Parallelogram	Opposite sides equal and parallel	Opposite angles equal	Bisect each other	0	2	$A = \text{base} \times \text{height}$ $A = bh$
Rhombus	4 equal sides, opposite sides parallel	Opposite angles equal	Bisect each other at 90°	2	2	$A = \text{base} \times \text{height}$ $A = bh$
Trapezium	1 pair of parallel sides					$A = \frac{1}{2}(a + b)h$
Isosceles trapezium	1 pair of parallel sides, non-parallel sides equal	2 pairs of equal angles	Equal in length	1	1*	$A = \frac{1}{2}(a + b)h$
Kite	2 pairs of adjacent sides equal	1 pair of opposite angles equal	One bisects the other at 90°	1	1*	

*A shape is only described as having rotational symmetry if the order of rotational symmetry is 2 or more.

- You should be able to use properties of quadrilaterals to solve problems.

Eg 1 Work out the size of the angle marked x.

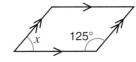

Opposite angles are equal.
So, $125° + 125° + x + x = 360°$
$x = 55°$

Eg 2 Find the area of this trapezium.

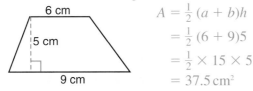

$A = \frac{1}{2}(a + b)h$
$= \frac{1}{2}(6 + 9)5$
$= \frac{1}{2} \times 15 \times 5$
$= 37.5\,\text{cm}^2$

- You should be able to construct a quadrilateral from given information using ruler, protractor, compasses.

1 This rectangle is drawn on 1 cm squared paper.
It has a perimeter of 18 cm.  **Not full size**
   (a)   What is the area of the rectangle?
   (b)   (i)   On 1 cm squared paper draw three different rectangles which each have a
            perimeter of 18 cm.
        (ii)   Find the area of each rectangle.

2 A quadrilateral with 4 equal sides and 4 right angles is called a square.
What is the mathematical name given to:
   (a)   A quadrilateral with 4 equal sides but no right angles?
   (b)   A quadrilateral with 2 pairs of opposite sides equal but diagonals of different lengths?
   (c)   A quadrilateral with only 1 pair of parallel sides of unequal lengths?   AQA

3 Find the size of the lettered angles.
   (a)   (b)   (c)

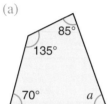 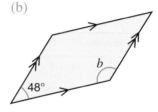 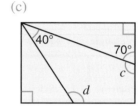

4 The diagram shows a quadrilateral ABCD.
   $AB = BC$ and $CD = DA$.
   (a)   Which of the following correctly describes the quadrilateral ABCD?
           rhombus   **parallelogram**   **kite**   **trapezium**
   (b)   Angle $ADC = 36°$ and angle $BCD = 105°$.
        Work out the size of angle ABC.

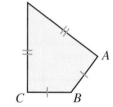

5 A rectangle is 4 cm wide and 9 cm long.
What is the length of the side of a square with exactly the same area?

          9 cm

                    4 cm

                                              AQA

6 A rectangle measures 8.6 cm by 6.4 cm.
   (a)   Find the perimeter of the rectangle.
   (b)   Find the area of the rectangle.

7 Two triangles are joined together to form a rhombus, as shown.
The perimeter of the rhombus is 36 cm.
The perimeter of each triangle is 24 cm.
Find the value of b.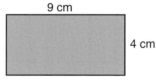

                                              AQA

8 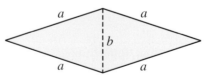 The diagram shows a trapezium PQRS.

   (a)   Work out the size of the angle marked x.

   (b)   Calculate the area of the trapezium.

Polygons

What you need to know

- A **polygon** is a many-sided shape made by straight lines.
- A polygon with all sides equal and all angles equal is called a **regular polygon**.
- Shapes you need to know:
  A 3-sided polygon is called a **triangle**.
  A 4-sided polygon is called a **quadrilateral**.
  A 5-sided polygon is called a **pentagon**.
  A 6-sided polygon is called a **hexagon**.
  An 8-sided polygon is called an **octagon**.
- The sum of the exterior angles of any polygon is 360°.
- At each vertex of a polygon: interior angle + exterior angle = 180°
- The sum of the interior angles of an n-sided polygon is given by:
  $(n - 2) \times 180°$
- For a regular n-sided polygon: exterior angle $= \dfrac{360°}{n}$

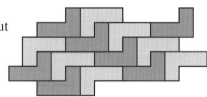

interior angle   exterior angle

- You should be able to use the properties of polygons to solve problems.

Eg 1 Find the sum of the interior angles of a pentagon.
$(5 - 2) \times 180° = 3 \times 180° = 540°$

> A pentagon has 5 sides, so, substitute
> $n = 5$ into $(n - 2) \times 180°$.

Eg 2 A regular polygon has an exterior angle of 30°.
(a)   How many sides has the polygon?
(b)   What is the size of an interior angle of the polygon?

(a) $n = \dfrac{360°}{\text{exterior angle}}$

$n = \dfrac{360°}{30°}$

$n = 12$

(b)   int. $\angle$ + ext. $\angle$ = 180°
int. $\angle$ + 30° = 180°
interior angle = 150°

- A shape will **tessellate** if it covers a surface without overlapping and leaves no gaps.
- All triangles tessellate.
- All quadrilaterals tessellate.
- Equilateral triangles, squares and hexagons can be used to make **regular tessellations**.

Exercise 25

1 Copy the shape onto squared paper.
On your diagram, draw **five** more shapes to show how the shape tessellates.

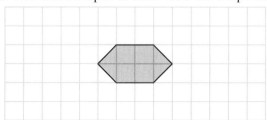

AQA

2 These shapes are drawn on isometric paper.

What are the differences between the symmetry of shape *A* and the symmetry of shape *B*?

3 Work out the size of the angles marked with letters.

(a)    (b) 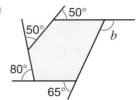   (c)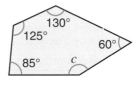

4 These shapes are regular polygons. Work out the size of the lettered angles.

(a) 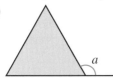   (b) 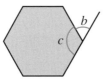   (c)

5 The diagram shows part of a regular polygon.
The exterior angles of this polygon are 24°.
How many sides has the polygon?

6 The diagram shows a hexagon.
Show that the sum of the interior angles of a hexagon is 720°.

7 *ABCDEFGHI* is a regular nonagon with centre *O*.

(a) Calculate the size of angle *x*.
(b) Calculate the size of angle *y*.

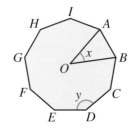

AQA

8 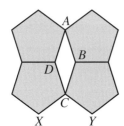 Four regular pentagons are placed together, as shown, to form a rhombus, *ABCD*.

Calculate the size of
(a) angle *ABC*,
(b) angle *XCY*.

9 *LM* and *MN* are two sides of a regular 10-sided polygon.
MNX is an isosceles triangle with *MX* = *XN*.
Angle *MXN* = 30°.

Work out the size of the obtuse angle *LMX*.

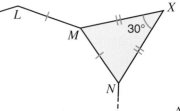

AQA

Direction and Distance

What you need to know

- **Compass points**

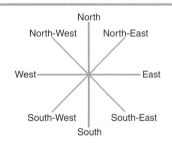

  The angle between North and East is 90°.
  The angle between North and North-East is 45°.

- **Bearings** are used to describe the direction in which you must travel to get from one place to another.

- A bearing is an angle measured from the North line in a clockwise direction.
  A bearing can be any angle from 0° to 360° and is written as a three-figure number.

  To find a bearing:
      measure angle *a* to find the bearing of *Y* from *X*,
      measure angle *b* to find the bearing of *X* from *Y*.

  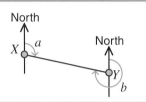

- You should be able to use **scales** and **bearings** to interpret and draw accurate diagrams.

  There are two ways to describe a scale.
  1. A scale of 1 cm to 10 km means that a distance of 1 cm on the map represents an actual distance of 10 km.
  2. A scale of 1 : 10 000 means that all distances measured on the map have to be multiplied by 10 000 to find the real distance.

Eg 1 The diagram shows the plan of a stage in a car rally.
The plan has been drawn to a scale of 1 : 50 000.

(a) What is the bearing of *Q* from *P*?
(b) What is the bearing of *P* from *R*?
(c) What is the actual distance from *P* to *R* in metres?

(a) 080°
(b) 295°
(c) 3500 m

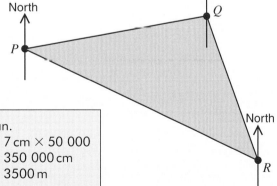

  PR is 7 cm on the plan.
  Actual distance *PR* = 7 cm × 50 000
                       = 350 000 cm
                       = 3500 m

Exercise 26

1 Jon is facing North-West.
He turns through 180°.
In which direction is he now facing?

2 This is a map of Scotland with five towns marked.

(a) Which town is South-West of Aberdeen?

(b) Write the direction South-West as a 3-figure bearing.

AQA

3 A bridge is 2600 m in length.
A plan of the bridge has been drawn to a scale of 1 cm to 100 m.
What is the length of the bridge on the plan?

4

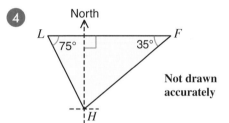

Not drawn accurately

A fishing boat sails from a harbour, *H*, to a point, *F*.
F is due East of a lighthouse, *L*.
Angle *FLH* is 75° and angle *LFH* is 35°.

(a) Calculate the bearing of *F* from *H*.
(b) Calculate the bearing of *L* from *H*.
(c) Calculate the bearing of *H* from *F*.

AQA

5 The map shows the position of a ship, *P*, and a lighthouse, *L*.

(a) What is the bearing of *P* from *L*?

Copy the diagram onto one-centimetre squared paper.
(b) Another ship, *Q*, is due North of *L*.
Q is on a bearing of 055° from *P*.
Mark clearly the position of *Q* on your diagram.

AQA

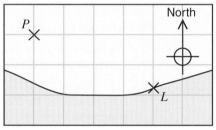

6

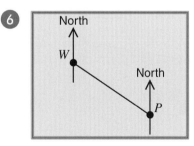

The map shows the positions of a windmill, *W*, and a pylon, *P*.

(a) What is the bearing of
(i) the pylon from the windmill,
(ii) the windmill from the pylon?

The map has been drawn to a scale of 2 cm to 5 km.
(b) Use the map to find the distance *WP* in kilometres.

7 The diagram shows a sketch of the course to be used for a running event.

(a) Draw an accurate plan of the course, using a
scale of 1 cm to represent 100 m.

(b) Use your plan to find
(i) the bearing of *X* from *Y*,
(ii) the distance *XY* in metres.

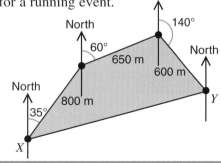

Circles

What you need to know

- A **circle** is the shape drawn by keeping a pencil the same distance from a fixed point on a piece of paper.
- Words associated with circles:

  Circumference – perimeter of a circle.

  Radius – distance from the centre of the circle to any point on the circumference. The plural of radius is **radii**.

  Diameter – distance right across the circle, passing through the centre point.

  Chord – a line joining two points on the circumference.

  Tangent – a line which touches the circumference of a circle at one point only. A tangent is perpendicular to the radius at the point of contact.

  Arc – part of the circumference of a circle.

  Segment – a chord divides a circle into two segments.

  Sector – two radii divide a circle into two sectors.

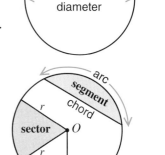

- The **circumference** of a circle is given by: $C = \pi \times d$ or $C = 2 \times \pi \times r$
- The **area** of a circle is given by: $A = \pi \times r^2$
- You should be able to solve problems which involve finding the circumference or the area of a circle.
- Take π to be 3.14 or use the π key on your calculator.

Eg 1   Calculate the circumference of a circle with diameter 18 cm.
Give your answer to 1 d.p.
$C = \pi \times d$
$C = \pi \times 18$
$C = 56.548\ldots$ $C = 56.5$ cm, correct to 1 d.p.

Eg 2   Find the area of a circle with radius 6 cm.
Give your answer to 3 sig. figs.
$A = \pi \times r^2$
$A = \pi \times 6 \times 6$
$A = 113.097\ldots$ $A = 113$ cm², correct to 3 sig. figs.

Eg 3   A circle has a circumference of 25.2 cm.
Find the diameter of the circle.
$C = \pi d$ so, $d = \dfrac{C}{\pi}$
$d = \dfrac{25.2}{\pi}$
$d = 8.021\ldots$ $d = 8.0$ cm, correct to 1 d.p.

Eg 4   A circle has an area of 154 cm².
Find the radius of the circle.
$A = \pi r^2$ so, $r^2 = \dfrac{A}{\pi}$
$r^2 = \dfrac{154}{\pi} = 49.019\ldots$
$r = \sqrt{49.019\ldots} = 7.001\ldots$ $r = 7$ cm, to the nearest cm.

Do not use a calculator for question 1.

1　A coin has a diameter of 1.96 cm.
Estimate the circumference of the coin.

1.96 cm

AQA

2　A circular pond has a radius of 3 metres.
　(a)　Calculate the circumference of the pond.
　(b)　Calculate the area of the pond.

AQA

3　A circle has a diameter of 7 cm.
　(a)　Calculate the circumference of this circle.
　(b)　Calculate the area of this circle.

AQA

4　Discs of card are used in the packaging of frozen pizzas.
Each disc fits the base of the pizza exactly.
Calculate the area of a disc used to pack a pizza.
Give your answer in terms of π.

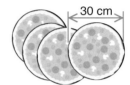

30 cm

5　　Tranter has completed three-fifths of a circular jigsaw puzzle.
The puzzle has a radius of 20 cm.
What area of the puzzle is complete?

6　Mr Kray's lawn is 25 m in length.
He rolls it with a garden roller.
The garden roller has a diameter of 0.4 m.
Work out the number of times the roller rotates when
rolling the length of the lawn once.

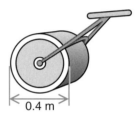

0.4 m

7　　The top of a table is a circle with a radius of 55 cm.
　(a)　Calculate the circumference of the table top.

On the table are 6 place mats.
Each place mat is a circle with a diameter of 18 cm.
　(b)　What area of the table top is **not** covered by place mats?

8　Calculate the area of a circle with radius 4.3 cm.

AQA

9　Each wheel on Hannah's bicycle has a radius of 15 cm.
Calculate how many complete revolutions each wheel makes when Hannah cycles 100 metres.

10　(a)　Jayne has a circular hoop of radius 35 cm.
　　　　Calculate the circumference of her hoop.
　(b)　Rashida has a hoop with a circumference of 300 cm.
　　　　Calculate the radius of Rashida's hoop.

AQA

11　Three circles overlap, as shown.
The largest circle has a diameter of 12 cm.
The ratio of the diameters $x : y$ is 1 : 2.
Calculate the shaded area.
Give your answer in terms of π.

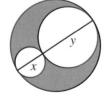

12　A circle has a circumference of 100 cm.
Calculate the area of the circle. Give your answer correct to three significant figures.

13　Alfie says, "A semi-circle with a radius of 10 cm has a larger area than a whole circle with
half the radius." Is he correct?
You **must** show working to justify your answer.

Areas and Volumes

What you need to know

- Shapes formed by joining different shapes together are called **compound shapes**.
  To find the area of a compound shape we must first divide the shape up into rectangles, triangles, circles, etc., and find the area of each part.
  Add the answers to find the total area.

  Eg 1 Find the total area of this shape.

  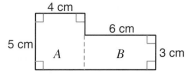

  $$\text{Area } A = 5 \times 4 = 20 \, \text{cm}^2$$
  $$\text{Area } B = 6 \times 3 = 18 \, \text{cm}^2$$
  $$\text{Total area} = 20 + 18 = 38 \, \text{cm}^2$$

- **Faces**, **vertices** (corners) and **edges**.

  Eg 2 A cube has 6 faces, 8 vertices and 12 edges.

  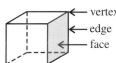

- A **net** can be used to make a solid shape.

  Eg 3 Draw a net of a cube.

- **Isometric paper** is used to make 2-D drawings of 3-D shapes.

  Eg 4 Draw a cube of edge 2 cm on isometric paper.

- **Plans and Elevations**
  The view of a 3-D shape looking from above is called a **plan**.
  The view of a 3-D shape from the front or sides is called an **elevation**.

  Eg 5 Draw diagrams to show the plan and elevation from **X**, for this 3-dimensional shape.

  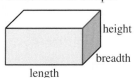

  plan   elevation X

  Dotted lines are used to show hidden edges.

- **Volume** is the amount of space occupied by a 3-dimensional shape.

- The formula for the volume of a **cuboid** is:
  Volume = length × breadth × height
  $$V = l \times b \times h$$

- Volume of a **cube** is: $V = l^3$

- To find the **surface area** of a cuboid, find the areas of the 6 rectangular faces and add the answers together.

  Eg 6 Find the volume and surface area of a cuboid measuring 7 cm by 5 cm by 3 cm.

  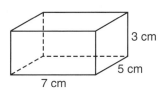

  $$\text{Volume} = l \times b \times h$$
  $$= 7 \, \text{cm} \times 5 \, \text{cm} \times 3 \, \text{cm}$$
  $$= 105 \, \text{cm}^3$$
  $$\text{Surface area} = (2 \times 7 \times 5) + (2 \times 5 \times 3) + (2 \times 3 \times 7)$$
  $$= 70 + 30 + 42$$
  $$= 142 \, \text{cm}^2$$

Eg 7 This cuboid has a volume of 75 cm³.
Calculate the height, h, of the cuboid.

$\text{Volume} = lbh$

$75 = 6 \times 5 \times h$

$h = \frac{75}{30}$

$h = 2.5\,\text{cm}$

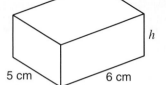

5 cm   6 cm

- **Prisms**
  If you make a cut at right angles to the length of a prism
  you will always get the same cross-section.

Triangular prism

cross-section

length

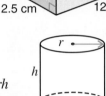

- Volume of a prism = area of cross-section × length

Eg 8 Calculate the volume of this prism.
The cross-section of this prism is a trapezium.

Area of cross-section $= \frac{1}{2}(5 + 3) \times 2.5$

$= 4 \times 2.5$

$= 10\,\text{cm}^2$

Volume of prism = area of cross-section × length

$= 10 \times 12$

$= 120\,\text{cm}^3$

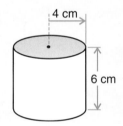

5 cm   3 cm

2.5 cm   12 cm

- A **cylinder** is a prism.
  Volume of a cylinder is: $\text{Volume} = \pi \times r^2 \times h$
  Surface area of a cylinder is: $\text{Surface area} = 2\pi r^2 + 2\pi rh$

h r

Eg 9 Calculate (a) the volume,
(b) the surface area of this cylinder.

4 cm

6 cm

(a) $\text{Volume} = \pi r^2 h$

$= \pi \times 4 \times 4 \times 6$

$= 301.592\ldots$

$= 302\,\text{cm}^3$, correct to 3 s.f.

(b) $\text{Surface area} = 2\pi r^2 + 2\pi rh$

$= 2 \times \pi \times 4 \times 4 + 2 \times \pi \times 4 \times 6$

$= 100.53\ldots + 150.796\ldots$

$= 251\,\text{cm}^2$, correct to 3 s.f.

Exercise 28   Do not use a calculator for question 1 to 5.

1 This shape is a pyramid.

(a) How many faces, edges and vertices has the pyramid?

(b) Which of these nets is a net of the pyramid?

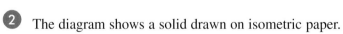

P   **Q**   **R**   **S**

2 The diagram shows a solid drawn on isometric paper.

(a) Draw the plan of the solid.

(b) Draw the elevation of the solid from
the direction shown by the arrow.

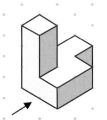

3 This shape has been drawn on 1 cm squared paper.

(a) Find the perimeter of the shape.

(b) Find the area of the shape.

(c) On 1 cm squared paper, draw a different shape with the same area.

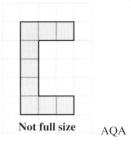

Not full size

AQA

4 Find the area of this shape.

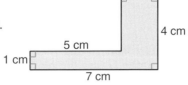

5 cm
4 cm
1 cm
7 cm

5 Identical blocks are used to make a base for a barbecue.

(a) Calculate the perimeter of the base.

(b) Calculate the length and the width of each block.

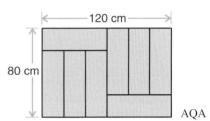

120 cm
80 cm

AQA

6

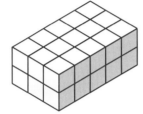

This cuboid has been made using cubes of side 1 cm.

(a) How many cubes are needed to make the cuboid?

(b) (i) Draw a net of the cuboid on 1 cm squared paper.
(ii) Hence, find the surface area of the cuboid.

7 (a) Which of these cuboids has the largest volume? Show all your working.

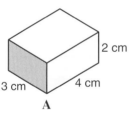
A
2 cm
3 cm
4 cm

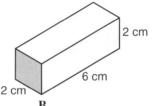
B
2 cm
6 cm
2 cm

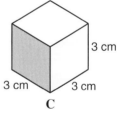
C
3 cm
3 cm
3 cm

(b) (i) Draw an accurate net of cuboid **A**.
(ii) Find the total surface area of cuboid **A**.

(c) Which cuboid has the largest surface area?

8

My cats eat two types of food.
Whiskat comes in a square-based tin of side 8.5 cm.
Yummy comes in a round-based tin of radius 5 cm.

(a) (i) Calculate the area of the base of the Whiskat tin.
(ii) Calculate the area of the base of the Yummy tin.
Give your answer correct to two decimal places.

(b) The heights of the two tins are the same.
Which tin holds more food when full?
Give a reason for your answer.

AQA

9 In the rectangle a triangular region has been shaded.
What percentage of the rectangle is shaded?
Give your answer to an appropriate degree of accuracy.

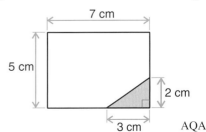

7 cm
5 cm
2 cm
3 cm

AQA

Areas and Volumes

10 A girder is 5 metres long.
Its cross-section is L-shaped, as shown.
Find the volume of the girder.

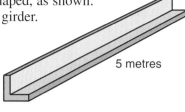

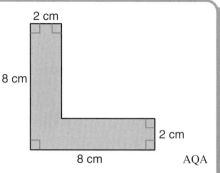

2 cm

8 cm

2 cm

8 cm

AQA

11

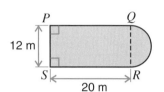

P   *Q*

12 m

S   *R*

20 m

The diagram shows the plan of a swimming pool.
The arc *QR* is a semi-circle.
PS = 12 m   and   *PQ* = *RS* = 20 m.
Calculate the area of the surface of the pool.

12 A triangular prism has dimensions, as shown.
(a)   Calculate the total surface area of the prism.
(b)   Calculate the volume of the prism.

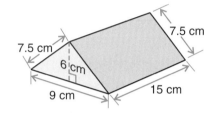

7.5 cm

7.5 cm

6 cm

9 cm

15 cm

13

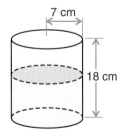

The diagram shows a solid.
Draw the elevation of this solid, from the direction shown by the arrow.

AQA

14 Ellie is cutting out circles of pastry to make jam tarts.
The radius of each circle is 3.5 cm.
She cuts 6 of these circles from a rectangular piece of
pastry 28 cm by 15 cm.
The pastry is 0.8 cm thick.
Calculate the volume of the pastry that is left.

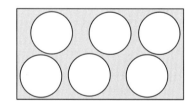

AQA

15 A cuboid has a volume of 100 cm³.
The cuboid is 8 cm long and 5 cm wide.
Calculate the height of the cuboid.

16

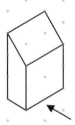

7 cm

18 cm

A cylinder of radius 7 cm and height 18 cm is half full of water.
One litre of water is added.
Will the water overflow?
You must show all your working.

AQA

17 A cylindrical water tank has radius 40 cm and height 90 cm.
(a)   Calculate the total surface area of the tank.

A full tank of water is used to fill a paddling pool.
(b)   The paddling pool is a square based prism, as shown.
Calculate the depth of water in the pool.

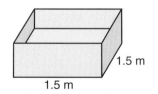

1.5 m

1.5 m

What you need to know

- The path of a point which moves according to a rule is called a **locus**.

- The word **loci** is used when we talk about more than one locus.

- You should be able to draw the locus of a point which moves according to a given rule.

  Eg 1 A ball is rolled along this zig-zag. Draw the locus of P, the centre of the ball, as it is rolled along.

- Using a ruler and compasses you should be able to carry out the **constructions** below.

  ❶ **The perpendicular bisector of a line.**

  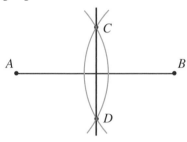

  Points on the line CD are **equidistant** from the points A and B.

  ❷ **The bisector of an angle.**

  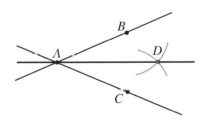

  Points on the line AD are **equidistant** from the lines AB and AC.

  ❸ **The perpendicular from a point to a line.**

  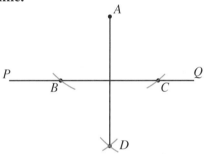

  ❹ **The perpendicular from a point on a line.**

  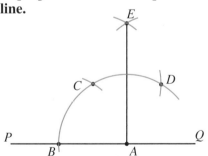

- You should be able to solve loci problems which involve using these constructions.

  Eg 2 P is a point inside triangle ABC such that:
  (i) P is equidistant from points A and B,
  (ii) P is equidistant from lines AB and BC.
  Find the position of P.

  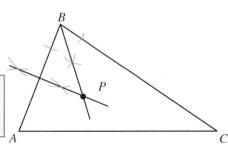

  > To find point P:
  > (i) construct the perpendicular bisector of line AB,
  > (ii) construct the bisector of angle ABC.

  P is at the point where these lines intersect.

1 The ball is rolled along the zig-zag.
Copy the diagram and draw the locus of the centre of the ball as it is rolled from X to Y.

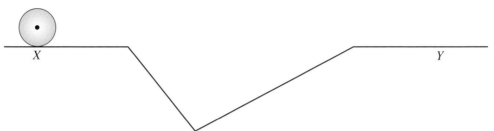

2 The diagram shows a plan of Paul's garden.
Draw the diagram using a scale of 1 cm to 1 m.

Paul has an electric lawnmower.
The lawnmower is plugged in at point P.

It can reach a maximum distance of 12 metres from P.
Using the same scale, show on your diagram the area
of the garden which the lawnmower can reach.

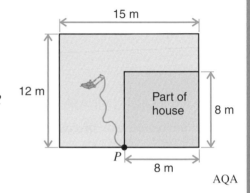

AQA

3 The map shows the positions of three villages A, B and C.
The map has been drawn to a scale of 1 cm to 2 km.

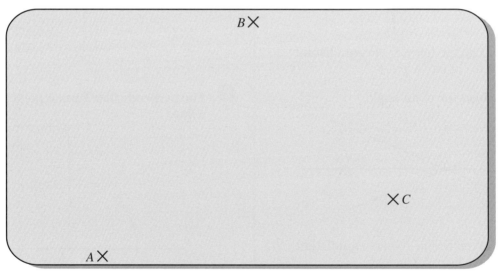

A supermarket is equidistant from villages A, B and C.
(a)   Copy the map and find the position of the supermarket on your diagram.
(b)   How many kilometres is the supermarket from village A?

4 (a)   Construct a kite $PQRS$ in which $PQ = PS = 7$ cm, $QR = RS = 5$ cm
and the diagonal $QS = 6$ cm.
X is a point inside the kite such that:
(i) X is equidistant from P and Q,
(ii) X is equidistant from sides PQ and PS.
(b)   By constructing the loci for (i) and (ii) find the position of X.
(c)   Measure the distance PX.

Transformations

What you need to know

- The movement of a shape from one position to another is called a **transformation**.

- **Single transformations** can be described in terms of a reflection, a rotation or a translation.

- **Reflection**: The image of the shape is the same distance from the mirror line as the original.

  Eg 1 Reflect shape *P* in the line *AB*.

  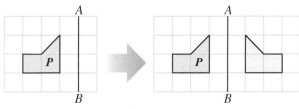

- **Rotation**: All points are turned through the same angle about the same point, called a centre of rotation.

  Eg 2 Rotate shape *P* 90° clockwise about the origin.

  Clockwise means:

  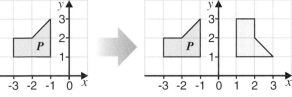

- **Translation**: All points are moved the same distance in the same direction without turning.

  Eg 3 Translate shape *P* with vector $\begin{pmatrix} 3 \\ 1 \end{pmatrix}$.

  $\begin{pmatrix} 3 \\ 1 \end{pmatrix}$ means 3 units right and 1 unit up.

  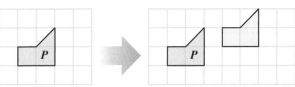

- You should be able to fully describe transformations.

Transformation	Image same shape and size?	Details needed to describe the transformation
Reflection	Yes	Mirror line, sometimes given as an equation.
Rotation	Yes	Centre of rotation, amount of turn, direction of turn.
Translation	Yes	Horizontal movement and vertical movement. Vector: top number = horizontal movement, bottom number = vertical movement.

Eg 4 Describe the single transformation which maps
  (a) *A* onto *B*,
  (b) *C* onto *A*,
  (c) *A* onto *D*.

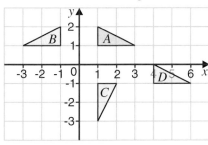

  (a) **Reflection** in the *y* axis.
  (b) **Rotation** of 90° anticlockwise about the origin.
  (c) **Translation** 3 units to the right and 2 units down.

1 Copy each diagram and draw the transformation given.

(a) Reflect the shape in the x axis.

(b) Translate the shape 2 units left and 3 units up.

(c) Rotate the shape 90° clockwise about the origin.

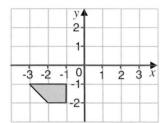

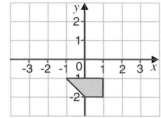

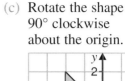

2 In each diagram A is mapped onto B by a single transformation. Describe each transformation.

(a)

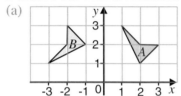

(b)

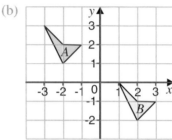

(c)

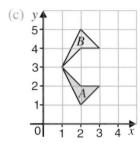

3 Copy the diagram and draw the reflection of the shaded triangle in the y axis.

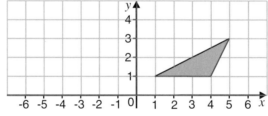

AQA

4 The diagram shows two positions of a shape.
Describe fully the single transformation which takes A onto B.

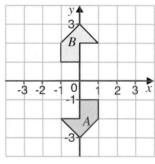

AQA

5 The diagram shows the positions of kites P, Q and R.

(a) P is mapped onto Q by a reflection.
What is the equation of the line of reflection?

(b) P is mapped onto R by a translation.
Describe the translation.

(c) P is mapped onto T by a rotation through 90° clockwise about (0, 3).
On squared paper, copy P and draw the position of T.

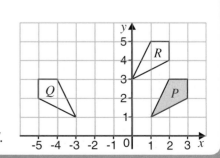

6

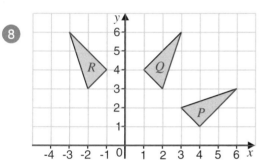

(a) Describe fully the **single** transformation which takes triangle *A* onto triangle *B*.

(b) Triangle *A* is rotated onto triangle *C*.
  (i) Write down the angle of rotation.
  (ii) Write down the coordinates of the centre of rotation.

(c) Triangle *C* is translated by the vector $\begin{pmatrix} 1 \\ -4 \end{pmatrix}$. Copy the grid and draw the new position of triangle *C*.

AQA

7 Describe the single transformation which maps

(a) *A* onto *B*,
(b) *A* onto *C*,
(c) *A* onto *D*.

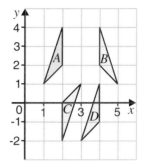

8

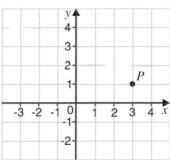

The diagram shows shapes *Q* and *R* which are transformations of shape *P*.
Describe fully the **single** transformation which takes

(a) *P* onto *R*,
(b) *P* onto *Q*.

AQA

9 Find the coordinates of the reflection of the point (1, 4) in the line $y = -x$. (You may find it useful to draw a diagram.)

AQA

10 The diagram shows the position of *P*.
P has coordinates (3, 1).

(a) *P* is mapped onto *Q* by a reflection in the line $y = x$. What are the coordinates of *Q*?

The translation $\begin{pmatrix} -2 \\ 1 \end{pmatrix}$ maps *P* onto *R*. The translation $\begin{pmatrix} 3 \\ -4 \end{pmatrix}$ maps *R* onto *S*.

(b) (i) What are the coordinates of *S*?
   (ii) What is the translation which maps *S* onto *P*?

(c) *T* has coordinates $(-1, 1)$.
   P is mapped onto *T* by a rotation through 90° anticlockwise about centre *X*.
   What are the coordinates of *X*?

AQA

Transformations . . . Transformations . . . Transformations . . .

Enlargements and Similar Figures

What you need to know

- When a shape is **enlarged**: all **lengths** are multiplied by a **scale factor**,
  angles remain unchanged.
  New length = scale factor × original length.

  The size of the original shape is:
  increased by using a scale factor greater than 1,
  reduced by using a scale factor which is a fraction, i.e. between 0 and 1.

- You should be able to draw an enlargement.

  Eg 1 Draw an enlargement of shape **P**, with scale factor 2, centre **O**.

  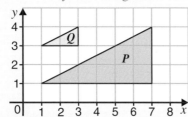

- You should be able to describe an enlargement.

  Eg 2 Describe fully the enlargement which maps **P** onto **Q**.

  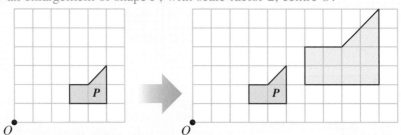

  Scale factor = $\dfrac{\text{new length}}{\text{original length}}$

  The centre of enlargement is the point where lines drawn through corresponding vertices of shapes **P** and **Q** cross.

  Enlargement, scale factor $\frac{1}{3}$, centre (1, 4).

- When two figures are **similar**:
  their **shapes** are the same, their **angles** are the same,
  corresponding **lengths** are in the same ratio, this ratio is the **scale factor** of the enlargement.

  Scale factor = $\dfrac{\text{new length}}{\text{original length}}$ New length = scale factor × original length

- All circles are similar to each other.

- All squares are similar to each other.

- You should be able to find corresponding lengths in similar shapes.

  Eg 3 These two shapes are similar.
  (a) Find the lengths of the sides marked x and y.
  (b) Find angle **PQR**.

  > AB and PQ are corresponding sides.
  > Scale factor = $\dfrac{PQ}{AB} = \dfrac{5}{3}$

  (a) $x = 4.5 \times \frac{5}{3} = 7.5\,\text{cm}$

  $y = 10 \div \frac{5}{3} = 6\,\text{cm}$

  (b) Angles stay the same.
  $\angle PQR = 100°$

  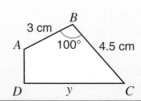

  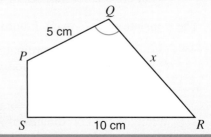

1 Copy the triangle onto squared paper and enlarge it by a
scale factor of 4.
Use the point marked *P* as the centre of enlargement.

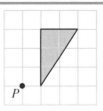

AQA

2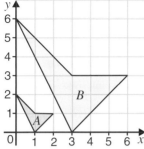

A is mapped onto *B* by a single transformation.
Describe the transformation.

3 (a) *P* is mapped onto *S* by an enlargement.
What is the centre and scale factor of the enlargement?
(b) Copy shape *P* onto squared paper.
Draw an enlargement of shape *P* with scale factor 2, centre (3, 2).

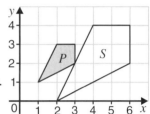

4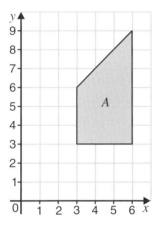

The diagram shows shape *A*.
Copy the diagram.
Draw the enlargement of shape *A* with scale factor $\frac{1}{3}$ and
centre of enlargement (0, 0).

AQA

5 The diagram shows rectangles **A**, **B** and **C**.

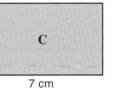

2 cm **A** 5 cm 3 cm **B** 4 cm **C** 7 cm

(a) Explain why rectangles **A** and **C** are **not** similar.
(b) Rectangles **A** and **B** are similar.
Work out the length of rectangle **B**.

6 These triangles are similar.
(a) Work out the length of the side *XY*.
(b) Work out the length of the side *AC*.

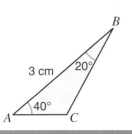

Pythagoras' Theorem

What you need to know

- The longest side in a right-angled triangle is called the **hypotenuse**.

- The **Theorem of Pythagoras** states:
  "In any right-angled triangle the square on the hypotenuse is equal to the sum of the squares on the other two sides."
  $$a^2 = b^2 + c^2$$

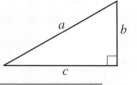

- When we know the lengths of two sides of a right-angled triangle, we can use the Theorem of Pythagoras to find the length of the third side.

$$a^2 = b^2 + c^2$$
Rearranging gives: $b^2 = a^2 - c^2$
$$c^2 = a^2 - b^2$$

Eg 1 Calculate the length of side a.

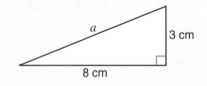

$a^2 = b^2 + c^2$
$a^2 = 8^2 + 3^2$
$a^2 = 64 + 9 = 73$
$a = \sqrt{73} = 8.544\ldots$
$a = 8.5\,\text{cm}$, correct to 1 d.p.

Eg 2 Calculate the length of side b.

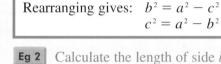

$b^2 = a^2 - c^2$
$b^2 = 9^2 - 7^2$
$b^2 = 81 - 49 = 32$
$b = \sqrt{32} = 5.656\ldots$
$b = 5.7\,\text{cm}$, correct to 1 d.p.

Exercise 32

Do not use a calculator for questions 1 and 2.

1 ABC is a right-angled triangle.
$AB = 5\,\text{cm}$ and $AC = 12\,\text{cm}$.
Calculate the length of BC.

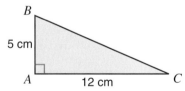

2 The positions of three villages, Oldacre (O), Adchester (A) and Byetoft (B),
are shown on the diagram.
Angle $OAB = 90°$.
The distance from Oldacre to Adchester is 8 km.
The distance from Oldacre to Byetoft is 10 km.
Calculate the distance from Adchester to Byetoft.

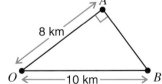

AQA

3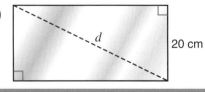

The diagram shows a rectangular sheet of paper.
The paper is 20 cm wide and the diagonal, d, is 35 cm.
Calculate the length of the sheet of paper.

4 Calculate the length of the line joining the points $A(-3, 2)$ and $B(6, -2)$.

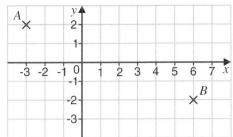

AQA

5

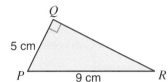

PQR is a right-angled triangle.
$PQ = 5\,\text{cm}$ and $PR = 9\,\text{cm}$.

Calculate the length of QR and,
hence, find the area of triangle PQR.

6 An oil rig is 15 km East and 12 km North from Kirrin.

(a) Calculate the direct distance from Kirrin to the oil rig.
(b) An engineer flew 14 km from Faxtown to the oil rig.
The oil rig is 10 km West of Faxtown.
Calculate how far South the oil rig is from Faxtown.

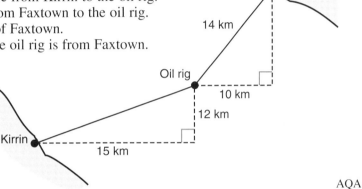

AQA

7 In the diagram, $BCDX$ is a square.

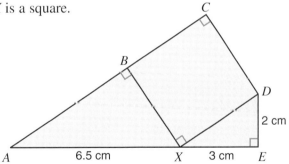

AXE is a straight line with $AX = 6.5\,\text{cm}$ and $XE = 3\,\text{cm}$. $DE = 2\,\text{cm}$.

(a) Calculate the area of $BCDX$.
(b) Calculate the length of AB, correct to one decimal place.

8

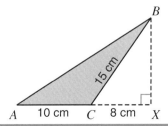

The diagram shows a triangle ABC.
Angle $BXA = 90°$, $BC = 15\,\text{cm}$,
$CX = 8\,\text{cm}$ and $AC = 10\,\text{cm}$.

Calculate the area of triangle ABC.
Give your answer correct to 3 significant figures.

AQA

Understanding and Using Measures

What you need to know

- The common units — both **metric** and **imperial** — used to measure **length**, **mass** and **capacity**.

- How to estimate measurements using sensible units and a suitable degree of accuracy.

- How to convert from one unit to another. This includes knowing the connection between one metric unit and another and the approximate equivalents between metric and imperial units.

Metric Units	Imperial Units	Conversions
Length 1 kilometre (km) = 1000 metres (m) 1 m = 100 centimetres (cm) 1 cm = 10 millimetres (mm) **Mass** 1 tonne (t) = 1000 kilograms (kg) 1 kg = 1000 grams (g) **Capacity and volume** 1 litre = 1000 millilitres (ml) 1 cm³ = 1 ml	**Length** 1 foot = 12 inches 1 yard = 3 feet **Mass** 1 pound = 16 ounces 14 pounds = 1 stone **Capacity and volume** 1 gallon = 8 pints	**Length** 5 miles is about 8 km 1 inch is about 2.5 cm 1 foot is about 30 cm **Mass** 1 kg is about 2.2 pounds **Capacity and volume** 1 litre is about 1.75 pints 1 gallon is about 4.5 litres

- How to change between units of area. For example $1\,m^2 = 10\,000\,cm^2$.

- How to change between units of volume. For example $1\,m^3 = 1\,000\,000\,cm^3$.

- You should be able to solve problems involving different units.

  Eg 1 A tank holds 6 gallons of water.
  How many litres is this? $6 \times 4.5 = 27$ litres

  Eg 2 A cuboid measures 1.5 m by 90 cm by 80 cm.
  Calculate the volume of the cuboid, in m³. $1.5 \times 0.9 \times 0.8 = 1.08\,m^3$

- Be able to read scales accurately.

  Eg 3 Part of a scale is shown.
  It measures weight in grams.
  What weight is shown by the arrow?

  The arrow shows 27 grams.

- Be able to recognise limitations on the accuracy of measurements.
  A measurement given to the nearest whole unit may be inaccurate by one half of a unit in either direction.

  Eg 4 A road is 400 m long, to the nearest 10 m.
  Between what lengths is the actual length of the road?
  Actual length = 400 m ± 5 m   395 m ≤ actual length < 405 m

- By analysing the **dimensions** of a formula it is possible to decide whether a given formula represents a **length** (dimension 1), an **area** (dimension 2) or a **volume** (dimension 3).

  Eg 5 p, q, r and s represent lengths.
  By using dimensions, decide whether the expression $pq + qr + rs$
  could represent a perimeter, an area or a volume.
  Writing $pq + qr + rs$ using dimensions:
  $$L \times L + L \times L + L \times L = L^2 + L^2 + L^2 = 3L^2$$
  So, $pq + qr + rs$ has dimension 2 and could represent an area.

1 Write down the metric unit you would use to measure
- (a) the length of a train,
- (b) the weight of a small pot of jam,
- (c) the amount of milk produced by a herd of cows each day.

2 A bookcase is 140 cm in height and weighs 8.5 kg.
- (a) What is the height of the bookcase in metres?
- (b) What is the weight of the bookcase in grams?

3 A glass contains 250 ml of milk. What fraction of a litre is this?

4 What value is shown by the pointer on each of these diagrams?

(a)

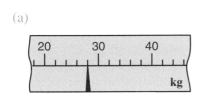

(b)

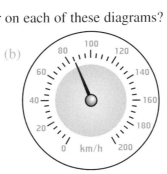

(c)

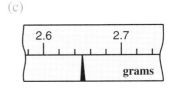

AQA

5 Write each of the following using a more suitable unit.
- (a) The distance between two towns is 6000 metres.
- (b) A mouse weighs 0.06 kilograms.
- (c) A piece of paper has an area of $0.006 \, m^2$.
- (d) A room has a volume of $60\,000\,000 \, cm^3$.

6 Two villages are 40 km apart.
- (a) Change 40 km into metres.
- (b) How many miles are the same as 40 km? AQA

7 (a) A towel measures 150 cm by 90 cm. Calculate the area of the towel in square metres.
- (b) Change $0.2 \, m^3$ to cm^3.

8 Serena measures the height of door A as 2 metres.
Tom measures the height of door B as 70 inches.
Which door is the higher, A or B? You must show all your working. AQA

9 Last year Felicity drove 2760 miles on business.
Her car does 38 miles per gallon. Petrol costs 89 pence per litre.
She is given a car allowance of 25 pence per kilometre.
How much of her car allowance is left after paying for her petrol?
Give your answer to the nearest £.

10 Mr Jones weighs his case on his bathroom scales which weigh to the nearest kilogram.
He finds that his case weighs 20 kg.
What are the greatest and least weights of the case? AQA

11 Jan cycles at 10 mph. Convert this speed to km/h. AQA

12 A bag of carrots weighs 2.5 kg, correct to the nearest 100 g.
What is the minimum weight of the bag of carrots?

13 a, b and c represents lengths.
Which of these formula could represent a volume?

$$\frac{abc}{2} \qquad a + b + c \qquad ab + bc + ca$$

Shape, Space and Measures Non-calculator Paper

Do not use a calculator for this exercise.

1 The diagram shows a line, *AB*, and a point, *C*.
Copy the diagram onto 1 cm squared paper.

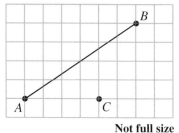

(a) Measure the length of the line *AB* in centimetres.
(b) Mark the midpoint of the line *AB* with a cross.
(c) Draw the line through *C* which is parallel to the line *AB*.

AQA

Not full size

2

A   B   C

The diagram shows some 3-dimensional shapes.

(a) How many edges has shape *A*?
(b) How many faces has shape *B*?
(c) What is the mathematical name for shape *C*?

3 (a) The shape has been drawn on 1 cm squared paper.
What is the area of the shape?

(b) This solid has been made using 1 cm cubes.

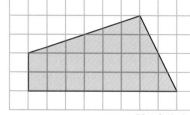

Not full size

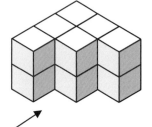

(i) What is the volume of the solid?
(ii) Draw the plan of the solid.
(iii) Draw the elevation of the solid from the
direction shown by the arrow.

4 (a) Use your protractor to measure the size of angles *x* and *y*.

(b) Which of these angles is an obtuse angle?

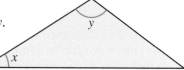

5 Find the angles marked with letters.
Give a reason for each of your answers.

(a)   (b)   (c)

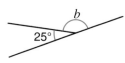

163°

47°

c

6 (a) In this triangle, all the sides are the same length.
(i) What name is given to this special type of triangle?

(ii) What name is given to this regular polygon?

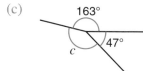

(b) This shape is made from three of the triangles and the polygon.
(i) Write down the order of rotational symmetry of the shape.
(ii) How many lines of symmetry has the shape?

AQA

7 (a) Describe the transformation which takes triangle *A* onto triangle *B*.

(b) *AFGHI* is an enlargement of the shaded shape *ABCDE*.

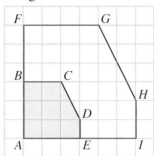

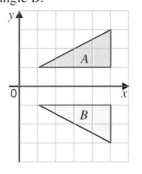

(i) What is the scale factor of the enlargement?
(ii) Write down two lines which are parallel.

AQA

8 A jar contains 400 grams of marmalade. Peter buys 12 jars of marmalade.
(a) How much marmalade, in total, is there in these 12 jars? Give your answer in kilograms.
(b) Does this total amount of marmalade weigh more or less than ten pounds?
You **must** show your working.

AQA

9 Find the size of the angles *a*, *b* and *c*. Give a reason for each of your answers.

(a) 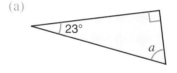 (b) 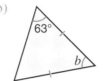 (c)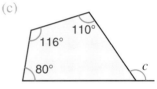

10 The descriptions of two different quadrilaterals are given in the boxes below.
Choose the correct names from this list.

kite   rectangle   rhombus   square   trapezium

(a) One pair of sides are parallel but they are not equal in length.

(b) The diagonals are of different lengths and cross at 90°.
Each diagonal is a line of symmetry of the quadrilateral.

AQA

11 Work out the area of each shape.

(a)  (b)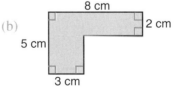

12 The diagram shows a cuboid.
By rounding each of the measurements to one significant figure,
estimate the volume of the cuboid.
You must show all your working.

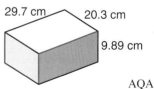

AQA

13 Use only the information given to find two triangles which are congruent to each other.
Give a reason for your answer.

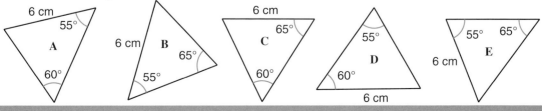

14 The sketch shows the positions of three footpaths which meet at A, B and C.
A is due north of C.
Triangle ABC is equilateral.

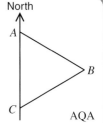

(a) Write down the three-figure bearing of B from C.
(b) Write down the three-figure bearing of A from B.

AQA

15 A circle of radius 5 cm is cut into quarters.
The quarters are put together to make shape S, as shown.

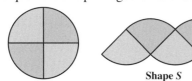

Shape S

(a) Calculate the area of the shape S.
Give your answer in terms of π.

(b) Calculate the perimeter of shape S.
Give your answer in terms of π.

(c) A different shape, T, is made from two of the quarter circles, of radius 5 cm, as shown.

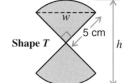

Shape T

(i) Calculate the width of shape T (marked w on the diagram).
Leave your answer as a square root.
(ii) State the height of shape T (marked h on the diagram).

AQA

16

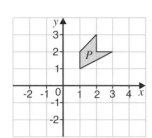

Copy the diagram onto squared paper.

(a) P is mapped onto Q by an enlargement, scale factor 2, centre $(-1, 3)$. Draw and label Q.

(b) P is mapped onto R by a translation with vector $\begin{pmatrix} -3 \\ 2 \end{pmatrix}$.
Draw and label R.

(c) P is mapped onto S by a rotation through $90°$ clockwise, about $(1, 0)$. Draw and label S.

17 Calculate the size of an interior angle of a regular octagon.

AQA

18 (a) Construct triangle ABC, in which $AB = 9.5$ cm, $BC = 8$ cm and $CA = 6$ cm.
(b) Using ruler and compasses only, bisect angle BAC.
(c) Shade the region inside the triangle where all the points are less than 7.5 cm from B, and nearer to AC than to AB.

19 These two cars are similar.
Calculate h, the height, of the smaller car.

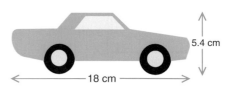

20 A circle has an area of 49π cm². Calculate the circumference of the circle in terms of π.

21 A helicopter is hovering 30 m above a boat. The distance is correct to the nearest metre.
What are the greatest and least distances of the helicopter from the boat?

22 The following formulae represent certain quantities connected with containers, where a, b and c are dimensions. πa abc $\sqrt{a^2 - c^2}$ $\pi a^2 b$ $2(a + b + c)$

(a) Explain why abc represents a volume.
(b) Which of these formulae represent lengths?

23 Calculate the volume of this triangular prism.

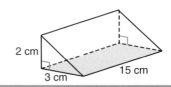

84

Shape, Space and Measures Calculator Paper

You may use a calculator for this exercise.

1

(a) Which of these weights are the same?
8000 g   80 kg   800 g   8 kg   0.08 kg

(b) Which of these lengths is the longest?
0.2 km   20 m   2000 mm   200 cm

(c) The scales show weights in kilograms.
Write down the weight of the pears.

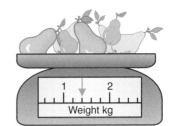

2

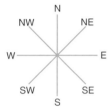

(a) What fraction of a complete turn is it from South to North-West?

(b) Isobel is facing East.

She makes a $\frac{1}{4}$ turn anticlockwise.

In which direction is she now facing?

3

(a) Draw a circle of radius 5 cm.
(b) Draw a diameter on your circle. Label its ends *A* and *B*.
(c) Mark a point, *P*, anywhere on the circumference of your circle.
Join *A* to *P* and *P* to *B*.
(d) Use your protractor to measure the angle *APB*.   AQA

4 Copy the diagram.
Draw the reflection of the shape in the mirror line.

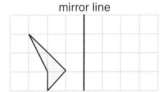
mirror line

5

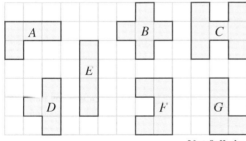

Not full size

Seven shapes are shown.
Each square on the grid has sides 1 cm long.

(a) What is the perimeter of shape *C*?
(b) Which shape is congruent to shape *A*?
(c) On 1 cm squared paper show how
shapes *B* and *F* may be used to make a
rectangle with an area of 15 cm².
You may use more than one of each shape.
AQA

6 The diagram shows points *A*, *B* and *C*.

(a) What are the coordinates of *A*?

(b) What are the coordinates of *C*?

(c) *ABCD* is a square.
What are the coordinates of *D*?

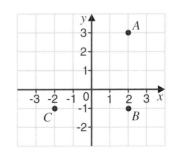

7

Copy the diagram onto squared paper.
Draw an enlargement of the kite, scale factor 2, centre *X*.

AQA

8 The diagram shows a sketch of a triangle.
  (a) Make an accurate drawing of the triangle.
  (b) Measure the size of angle PQR.
  (c) What type of angle is angle PQR?

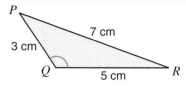

9 Colin is 5 feet 10 inches tall and weighs 11 stones.
On a medical form he is asked to give his height in centimetres and his weight in kilograms.
What values should he give?

10 The diagram shows the positions of shapes P, Q, R and S.

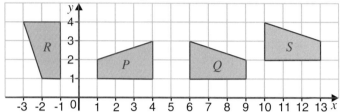

Describe the single transformation which takes:
(a) P onto Q,   (b) P onto R,   (c) Q onto S.

11 Find the size of the angles a, b, c and d.
  (a)   (b)

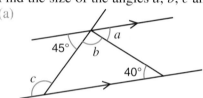 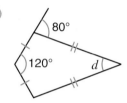

12 The length of a rectangle is 10.8 cm.
The perimeter of the rectangle is 28.8 cm.
Calculate the width of the rectangle.

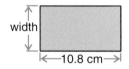

AQA

13
  (a) Copy the diagram.
      Shade two more squares so that the final diagram has line symmetry only.
  (b) Make another copy of the diagram.
      Shade two more squares so that the final diagram has
      rotational symmetry only.

14 (a) Part of a tessellation of triangles is shown.
      Copy the diagram.
      Continue the tessellation by drawing four more triangles.
  (b) Do all regular polygons tessellate?
      Give a reason for your answer.

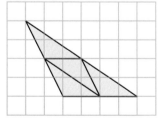

15 (a) A cuboid measures 2 cm by 2.5 cm by 4 cm.
      (i) Draw an accurate net of the cuboid.
      (ii) Calculate the total surface area of the cuboid.
  (b) Another cuboid has a volume of 50 cm³. The base of the cuboid measures 4 cm by 5 cm.
      Calculate the height of the cuboid.

16 A circular dish has a diameter of 9 cm. Calculate the circumference of the dish.   AQA

17 The area of the trapezium is 20 m².
The parallel sides a and b are different lengths.
The perpendicular height, h, is 4 m.
Find a possible pair of values for a and b.

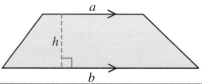

AQA

18 The diagram shows the angle formed when three regular polygons are placed together, as shown.

(a) Explain why angle a is 120°.

(b) Work out the size of the angle marked b.

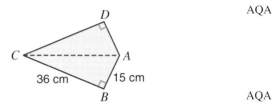

19 On a map the distance between two hospitals is 14.5 cm.
The map has been drawn to a scale of 1 to 250 000.
Calculate the actual distance between the hospitals in kilometres.

20

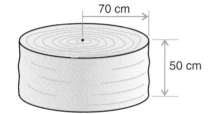

The diagram shows the points $P(0, -4)$ and $Q(5, 2)$.
Find the coordinates of the midpoint of the line segment PQ.

AQA

21 The diagram shows a kite $ABCD$.
(a) Calculate the area of the kite $ABCD$.
(b) Calculate the length of AC.

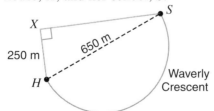

AQA

22

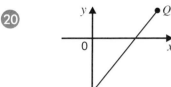

The diagram shows a bale of straw.
The bale is a cylinder with radius 70 cm and height 50 cm.
Calculate the volume of the bale.

AQA

23 Three oil rigs, X, Y and Z, are supplied by boats from port P.
X is 15 km from P on a bearing of 050°.
Y is 20 km from P on a bearing of 110°.
Z is equidistant from X and Y and 30 km from P.
(a) By using a scale of 1 cm to represent 5 km, draw an accurate diagram to show the positions of P, X, Y and Z.
(b) Use your diagram to find (i) the bearing of Y from Z,
(ii) the distance, in kilometres, of Y from Z.

24 The diagram shows Fay's house, H, and her school, S.

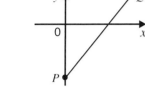

To get to school Fay has a choice of two routes.
She can either walk along Waverly Crescent or along the footpaths HX and XS.
Waverly Crescent is a semi-circle with diameter 650 m.
The footpath HX is 250 m and meets the footpath XS at right-angles.
Which of these routes is shorter? By how much?

25 A letter L is drawn, as shown. Draw the letter L accurately.
A point P is 2 cm from the letter L.
Draw the locus of all the possible positions of P.

AQA

Collection and Organisation of Data

What you need to know

- **Primary data** is data collected by an individual or organisation to use for a particular purpose. Primary data is obtained from experiments, investigations, surveys and by using questionnaires.

- **Secondary data** is data which is already available or has been collected by someone else for a different purpose.
  Sources of secondary data include the Annual Abstract of Statistics, Social Trends and the Internet.

- **Qualitative** data – Data which can only be described in words.   E.g. Colour of cars.

- **Quantitative** data – Data that has a numerical value.
  Quantitative data is either **discrete** or **continuous**.
  Discrete data can only take certain values.   E.g. Numbers of cars in car parks.
  Continuous data has no exact value and is measurable.   E.g. Weights of cars.

- **Data Collection Sheets** – Used to record data during a survey.

- **Tally** – A way of recording each item of data on a data collection sheet.
       A group of five is recorded as ⊬⊩⊩.

- **Frequency Table** – A way of collating the information recorded on a data collection sheet.

- **Grouped Frequency Table** – Used for continuous data or for discrete data when a lot of data has to be recorded.

- **Database** – A collection of data.

- **Class Interval** – The width of the groups used in a grouped frequency distribution.

- **Questionnaire** – A set of questions used to collect data for a survey.
  Questionnaires should:
     (1) use simple language,
     (2) ask short questions which can be answered precisely,
     (3) provide tick boxes,
     (4) avoid open-ended questions,
     (5) avoid leading questions,
     (6) ask questions in a logical order.

- **Hypothesis** – A hypothesis is a statement which may or may not be true.

- When information is required about a large group of people it is not always possible to survey everyone and only a **sample** may be asked.
  The sample chosen should be large enough to make the results meaningful and representative of the whole group (population) or the results may be **biased**.

- **Two-way Tables** – A way of illustrating two features of a survey.

Exercise 34

1   Harry wants to find out how people travel to work.
   (a)   (i)   Design an observation sheet for Harry to record data.
           (ii)   Complete your observation sheet by inventing data for 20 people.
   (b)   Harry decides to stand outside the bus station to collect his data.
         Give a reason why this is not a suitable place to carry out the survey.

2 The table shows information about pupils in the same class at a school.

Name	Gender	Month of birth	Day of birth
Corrin	F	June	Monday
Daniel	M	March	Thursday
Laila	F	May	Friday
Ria	F	March	Tuesday
Miles	M	April	Tuesday

(a) Who was born in May?

(b) Who was born on a Tuesday in March?

(c) Which of these pupils is most likely to be the youngest?
Give a reason for your answer.

3 Tayfan is organising a skiing holiday to Italy for his friends.
They can go to Cervinia, Livigno or Tonale.
He asks each of his friends which resort they would like to go to and records the answers in his notebook.

Cervinia	Cervinia	Livigno	Tonale
Tonale	Tonale	Livigno	Cervinia
Livigno	Cervinia	Tonale	Tonale
Cervinia	Livigno	Tonale	Livigno
Tonale	Cervinia	Livigno	Tonale
Livigno	Tonale	Cervinia	

Show a better way of recording this information.

4 Meeta is doing a survey about sport.
She asks the question,

"Do you play football, rugby or hockey?"

(a) Give a reason why this is not a suitable question.

(b) Write a similar question which is suitable.

5 Pat is investigating how long students spend on homework each night.
The time, in minutes, taken by 30 students to do their homework on a Wednesday night is shown.

100	55	45	80	65	40	10	45	105	60
35	40	30	45	90	25	120	55	60	75
70	45	90	45	90	45	25	15	20	75

(a) Using equal class intervals, copy and complete the frequency table to show this data.

Time (t minutes)	Tally	Frequency
$0 \leqslant t < 30$		

(b) Which class interval has the highest frequency?

(c) Give two reasons why this data may not be typical for these students.

6 Jane conducts a survey of the favourite colours of the students in her class.
She records the results.

Male	Red	Female	Green	Male	Green
Male	Yellow	Female	Green	Male	Green
Male	Red	Male	Red	Female	Yellow
Female	Green	Male	Yellow	Female	Red
Female	Red				

Record the results in a two-way table.

AQA

7 Wentbridge College wants to do a survey. This is one of the questions.

Do you agree that Wentbridge College provides better facilities than Mill Dam College?

(a) Give one criticism of this question.

(b) The survey is only carried out with students at Wentbridge College.
Give a reason why this is **not** suitable.

8 Jamie is investigating the use made of his college library. Here is part of his questionnaire:

Library Questionnaire

1. How old are you? ....

(a) (i) Give a reason why this question is unsuitable.

(ii) Rewrite the question so that it could be included.

(b) Jamie asks the librarian to give the questionnaires to students when they borrow books.

(i) Give reasons why this sample may be biased.

(ii) Suggest a better way of giving out the questionnaires.

9 The table shows the results of a survey of 500 people.

	Can drive	Cannot drive
Men	180	20
Women	240	60

A newspaper headline states: **Survey shows that more women can drive than men.**
Do the results of the survey support this headline?
Give a reason for your answer.

10 This sample was used to investigate the claim: **"Women do more exercise than men."**

	Age (years)			
	16 to 21	22 to 45	46 to 65	Over 65
Male	5	5	13	7
Female	25	35	0	0

Give three reasons why the sample is biased.

11 A mobile phone company wants to build a transmitter mast on land belonging to a school.
The company offers the school £50 000 for the land.
The local paper receives 20 letters objecting to the proposal and 5 letters in favour.
One of the paper's reporters writes an article in which he claims:

'Objectors outnumber those in favour by 4 to 1'

Give **two** reasons why the newspaper reporter's claim may **not** be correct.

12 The two-way table shows the number of credit cards and the number of store cards owned by each of 50 shoppers.

Number of store cards

		0	1	2	3
Number of credit cards	0	3	2	1	0
	1	5	4	3	1
	2	8	6	4	3
	3	4	3	2	1

(a) How many of the shoppers had two credit cards and one store card?

(b) How many of the shoppers had three credit cards?

(c) How many of the shoppers had exactly one card?

(d) How many of the shoppers had more credit cards than store cards?

Pictograms and Bar Charts

What you need to know

- **Pictogram**. Symbols are used to represent information.
  Each symbol can represent one or more items of data.

  Eg 1 A sports club has 45 members.
  Last Saturday, 15 played football, 13 played hockey and 17 played rugby.

  Draw a pictogram to show this information. Use ⚲ = 5 members.

Football	⚲ ⚲ ⚲
Hockey	⚲ ⚲ ⚲
Rugby	⚲ ⚲ ⚲ ⚲

  Note: ⚲ represents 3 members

  ⚲ represents 2 members

- **Bar chart**. Used for data which can be counted.
  Often used to compare quantities of data in a distribution.
  The length of each bar represents frequency.
  The longest bar represents the **mode**.
  The difference between the largest and smallest variable
  is called the **range**.

  > Bars can be drawn horizontally
  > or vertically.
  > Bars are the same width and
  > there are gaps between bars.

- **Bar-line graph**. Instead of drawing bars, horizontal or vertical lines are drawn to show
  frequency.

  Eg 2 The graph shows the number of goals scored by a football team in 10 matches.

  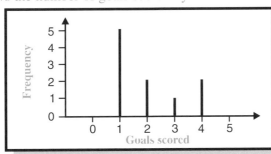

  (a) Which number of goals scored is the mode?
  (b) What is the range of the number of goals scored?

  (a) The tallest bar represents the mode. The mode is 1 goal.
  (b) The range is the difference between the largest and smallest number of goals scored.
  The range = 4 − 1 = 3

Exercise 35

1. A sample of retired people was asked, "Which television channel do you watch the most?"
   The table shows the results.

Television channel	BBC 1	BBC 2	ITV 1	Channel 4	Channel 5
Number of people	16	9	11	10	4

   (a) Draw a bar chart to show these results.
   (b) What percentage watched Channel 4 the most?

2 The pictogram shows the number of DVDs owned by each of four friends.

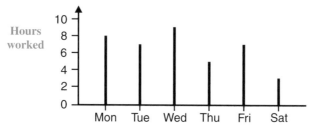

Gerry	⊙ ⊙ ◖	
Jack	⊙ ⊙ ⊙ ◖	⊙ = 10 DVDs
Tom	⊙ ⊙	
Harry	⊙	

(a) Who owns the most DVDs?
(b) How many more DVDs does Gerry own than Harry?

AQA

3 Philip asks his friends what their favourite sport is.
The results are shown in the tally chart.

(a) How many friends chose football?
(b) How many friends did Philip ask?
(c) Draw a pictogram to show Philip's results.

Use the symbol ⊼ to represent 4 friends.

Sport	Tally
Football	⟋⟋⟋ ⟋⟋⟋ ‖
Rugby	‖‖
Racing	⟋⟋⟋ ‖‖
Other	⟋⟋⟋ ⟋⟋⟋

AQA

4 The bar-line graph shows the number of hours a plumber worked each day last week.

(a) On which day did he work the most hours?
(b) How many more hours did he work on Monday than on Thursday?
(c) How many hours did he work altogether last week?

5 Causeway Hockey Club have a hockey team for men and a hockey team for women.
The bar chart shows the number of goals scored in matches played by these teams last season.

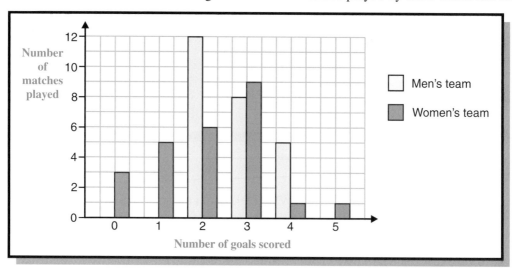

(a) For the men's team, find the range and mode in the number of goals scored.
(b) Compare and comment on the goals scored by these teams last season.

What you need to know

- There are three types of **average**: the **mode**, the **median** and the **mean**.
  The **mode** is the most common value.
  The **median** is the middle value (or the mean of the two middle values) when the values are arranged in order of size.

  The **Mean** $= \dfrac{\text{Total of all values}}{\text{Number of values}}$

- The **range** is a measure of **spread**, and is the difference between the highest and lowest values.

  Eg 1 The number of text messages received by 7 students on Saturday is shown.

  $$2 \quad 4 \quad 3 \quad 4 \quad 4 \quad 3 \quad 2$$

  Find (a) the mode, (b) the median, (c) the mean, (d) the range.

  (a) The mode is 4.

  (b) 2   2   3   ③   4   4   4   The median is 3.

  (c) The mean $= \dfrac{2 + 4 + 3 + 4 + 4 + 3 + 2}{7} = \dfrac{22}{7} = 3.14\ldots = 3.1$, correct to 1 d.p.

  (d) The range $= 4 - 2 = 2$

- To find the mean of a **frequency distribution** use:

  $$\text{Mean} = \frac{\text{Total of all values}}{\text{Number of values}} = \frac{\Sigma fx}{\Sigma f}$$

  Eg 2 The table shows the number of stamps on some parcels.

Number of stamps	1	2	3	4
Number of parcels	5	6	9	4

  Find the mean number of stamps per parcel.

  $$\text{Mean} = \frac{\text{Total number of stamps}}{\text{Number of parcels}} = \frac{1 \times 5 + 2 \times 6 + 3 \times 9 + 4 \times 4}{5 + 6 + 9 + 4} = \frac{60}{24} = 2.5$$

- To find the mean of a **grouped frequency distribution**, first find the value of the midpoint of each class.
  Then use:

  $$\text{Estimated mean} = \frac{\text{Total of all values}}{\text{Number of values}} = \frac{\Sigma fx}{\Sigma f}$$

  Eg 3 The table shows the weights of some parcels.

Weight (w grams)	Frequency
$100 \leqslant w < 200$	7
$200 \leqslant w < 300$	11
$300 \leqslant w < 400$	19
$400 \leqslant w < 500$	3

  Calculate an estimate of the mean weight of these parcels.

  $$\text{Mean} = \frac{\Sigma fx}{\Sigma f} = \frac{150 \times 7 + 250 \times 11 + 350 \times 19 + 450 \times 3}{7 + 11 + 19 + 3} = \frac{11\,800}{40} = 295 \text{ grams}$$

● You should be able to choose the best average to use in different situations:
   When the most **popular** value is wanted use the **mode**.
   When **half** of the values have to be above the average use the **median**.
   When a **typical** value is wanted use either the **mode** or the **median**.
   When all the **actual** values have to be taken into account use the **mean**.
   When the average should not be distorted by a few very small or very large values
   do **not** use the mean.

Exercise 36
Do not use a calculator for questions 1 to 3.

1 Nine students were asked to estimate the length of this line, correct to the nearest centimetre.

The estimates the students made are shown.
   8   10   10   10   11   12   12   14   15
   (a)   What is the range in their estimates?
   (b)   Which estimate is the mode?
   (c)   Which estimate is the median?
   (d)   Work out the mean of their estimates.

2 The prices paid for eight different meals at a restaurant are:
   £10   £9   £9.50   £12   £20   £11.50   £11   £9
   (a)   Which price is the mode?
   (b)   Find the median price.
   (c)   Calculate the mean price.
   (d)   Which of these averages best describes the average price paid for a meal?
          Give a reason for your answer.

3 (a)   Calculate the mean of 13.9, 15.3, 11.7 and 16.2.
   (b)   Using your result from part (a), explain how to find quickly the mean of
                14.9, 16.3, 12.7 and 17.2
   (c)   Calculate the median of the numbers in part (a).
   (d)   If the number 16.2 in part (a) was changed to 27.2, explain, without doing a calculation,
          whether the mean or the median would be more affected.
                                                                                    AQA

4 A company puts this advert in the local paper.

> AQA Motor Company
> **MECHANIC NEEDED**
> *Average wage over £400 per week*

The following people work for the company.

Job	Wages per week (£)
Apprentice	200
Cleaner	200
Foreman	350
Manager	800
Mechanic	250
Parts Manager	520
Sales Manager	620

   (a)   What is the mode of these wages?
   (b)   What is the median wage?
   (c)   Calculate the mean wage.
   (d)   Explain why the advert is misleading.
                                                                                    AQA

5 (a) The number of hours of sunshine each day last week is shown.

Monday	Tuesday	Wednesday	Thursday	Friday	Saturday	Sunday
5.3	6.4	3.7	4.8	7.5	8.6	5.7

   (i) What is the range in the number of hours of sunshine each day?
   (ii) Work out the mean number of hours of sunshine each day.

 (b) In the same week last year, the range in the number of hours of sunshine each day was 9 hours and the mean was 3.5 hours.
   Compare the number of hours of sunshine each day in these two weeks.

6 The graph shows the distribution of goals scored by a football team in home and away matches.

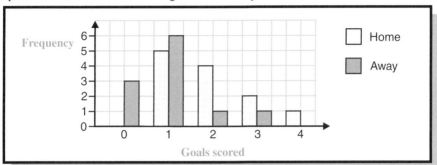

 (a) What is the range of the number of goals scored at home matches?
 (b) Calculate the mean number of goals per match for home matches.
 (c) A supporter says,

   > *"The average number of goals per match is the same for both away matches and home matches."*

   Which average is being used?   AQA

7 Four taxi drivers recorded how many passengers they carried on each journey during one evening. The table shows the numbers of journeys they made with different numbers of passengers.

		Number of passengers carried			
		1	2	3	4
Taxi	**A**	6	6	4	0
	B	7	7	3	1
	C	5	7	2	0
	D	4	4	3	1

 (a) Which taxi completed the most journeys that evening?
 (b) Calculate the total number of journeys in which exactly 3 passengers were carried.
 (c) There were 60 journeys made altogether.
   Calculate the mean number of passengers per taxi journey.   AQA

8 Fred records the time taken by 30 pupils to complete a cross-country run.

Time (t minutes)	Number of pupils
$20 \leqslant t < 25$	9
$25 \leqslant t < 30$	8
$30 \leqslant t < 35$	5
$35 \leqslant t < 40$	2
$40 \leqslant t < 45$	6

 (a) Calculate an estimate of the mean time taken to complete the run.
 (b) Which time interval contains the median time taken to complete the run? AQA

Pie Charts and Stem and Leaf Diagrams

What you need to know

- **Pie chart**. Used for data which can be counted.
  Often used to compare proportions of data, usually with the total.
  The whole circle represents all the data.
  The size of each sector represents the frequency of data in that sector.
  The largest sector represents the **mode**.

  Eg 1 The pie chart shows the makes of 120 cars.
  - (a) Which make of car is the mode?
  - (b) How many of the cars are Ford?

  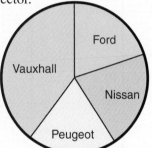

  - (a) The sector representing Vauxhall is the largest.
    Therefore, Vauxhall is the mode.
  - (b) The angle of the sector representing Ford is 72°.
    The number of Ford cars = $\frac{72}{360} \times 120 = 24$

- **Stem and leaf diagrams**. Used to represent data in its original form.
  Data is split into two parts.
  The part with the higher place value is the stem.   E.g.   15 = stem 1, leaf 5.
  A key is given to show the value of the data.   E.g.   3|4 means 3.4, etc.
  The data is shown in numerical order on the diagram.   E.g.   2|3 5 9 represents 23, 25, 29.

  Back to back stem and leaf diagrams can be used to compare two sets of data.

  Eg 2 The times, in seconds, taken by 10 students to complete a puzzle are shown.

  | 9 | 23 | 17 | 20 | 12 | 11 | 24 | 12 | 10 | 26 |

  Construct a stem and leaf diagram to represent this information.

  2 | 0 means 20 seconds

0	9				
1	0	1	2	2	7
2	0	3	4	6	

Exercise 37

1   The stem and leaf diagram shows the highest November temperature recorded in 12 European
countries last year.

0 | 7   means 7°C

0	7	9					
1	0	3	4	4	4	7	8
2	0	1	2				

- (a)   How many countries are included?
- (b)   What is the maximum temperature recorded?
- (c)   Which temperature is the mode?
- (d)   When the temperature in another European country is included in the data, the range
  increases by 2°C.
  What was the temperature in that country?
  Explain your answer.

2 Louise asks the children in her year group what type of house they live in. The results are shown in the pie chart.

(a) There are 12 children who live in detached houses.
How many children live in semi-detached houses?

(b) Calculate how many children Louise asked.

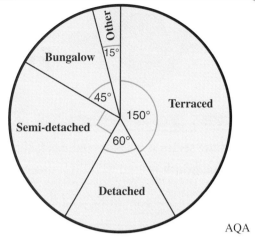

AQA

3 (a) The table shows the fuels used for heating in all the houses in a large town.

Fuel	Solid fuel	Electricity	Gas	TOTAL
Number of houses (in 1000's)	8	42	70	120

Draw a clearly labelled pie chart to represent this information.

(b) This pie chart shows the fuels used for heating in all the houses in a small village.
  (i) What fraction of these houses use gas?
  (ii) Solid fuel is used in 24 houses.
      How many houses are in the village?

(c) Use the information from these two pie charts to say which fuel is most likely to be the mode. Give a reason for your answer.

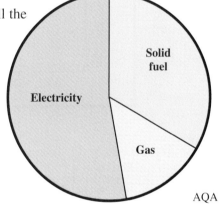

AQA

4 The number of text messages Anila sent each day in the last two weeks is shown.

   7   12   10   5   21   11   9   2   17   3   5   13   20   15

(a) Construct a stem and leaf diagram to show this information.
(b) What is the range in the number of text messages Anila sent each day?

5 The holiday destinations of 30 students were recorded.

Destination	France	Spain	Italy	Greece	America
Number of students	6	9	3	7	5

(a) Draw a clearly labelled pie chart to represent this information.
(b) What percentage of the 30 students went to France for their holiday?

AQA

6 Twenty children were asked to estimate the length of a leaf.
Their estimates, in centimetres, are:

Boys									
4.5	5.0	4.0	3.5	4.0	4.5	5.0	4.5	3.5	4.5

Girls									
4.5	5.0	3.5	4.0	5.5	3.5	4.5	3.5	3.0	2.5

(a) Construct a back to back stem and leaf diagram to represent this information.
(b) Compare and comment on the estimates of these boys and girls.

Time Series and Frequency Diagrams

What you need to know

- A **time series** is a set of readings taken at time intervals.
- A **line graph** is used to show a time series.

  Eg 1 The table shows the temperature of a patient taken every half-hour.

Time	0930	1000	1030	1100	1130	1200
Temperature °C	36.9	37.1	37.6	37.2	36.5	37.0

  (a) Draw a line graph to illustrate the data.
  (b) Estimate the patient's temperature at 1115.

  (a)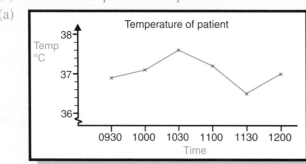

  To draw a line graph:
  Plot the given values.
  Points are joined by lines to show the **trend**.

  Only the plotted points represent **actual values**.
  The lines show the **trend** and can be used to **estimate values**.

  (b) 36.8°C

- **Histogram**. Used to illustrate **grouped frequency distributions.**
  The horizontal axis is a continuous scale.

- **Frequency polygon**. Used to illustrate grouped frequency distributions.
  Often used to compare two or more distributions on the same diagram.
  Frequencies are plotted at the midpoints of the class intervals and joined with straight lines.
  The horizontal axis is a continuous scale.

  Eg 2 The frequency distribution of the heights of some boys is shown.

Height (h cm)	$130 \leqslant h < 140$	$140 \leqslant h < 150$	$150 \leqslant h < 160$	$160 \leqslant h < 170$	$170 \leqslant h < 180$
Frequency	1	7	12	9	3

  Draw a histogram and a frequency polygon to illustrate the data.

  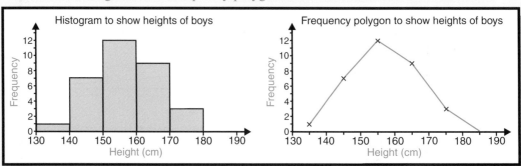

- **Misleading graphs**. Graphs may be misleading if:
  the scales are not labelled,   the scales are not uniform,   the frequency does not begin at zero.

1 On Sunday, Alfie records the outside temperature every two hours.
The temperatures he recorded are shown in the table.

Time of day	0800	1000	1200	1400	1600	1800
Outside temperature (°C)	9	12	15	17	16	14

(a) Draw a line graph to represent the data.
(b) What is the range in the temperatures recorded?
(c) (i) Use your graph to estimate the temperature at 1300.
(ii) Explain why your answer in (c)(i) is an estimate.

2 The amount of time spent by a group of pupils on their mobile phones in one week is recorded.
Here are the results.

Time (minutes)	Number of pupils
Less than 10 minutes	12
10 minutes or more but less than 20 minutes	9
20 minutes or more but less than 30 minutes	13
30 minutes or more but less than 40 minutes	6
40 minutes or more but less than 50 minutes	8
50 minutes or more but less than 60 minutes	2

(a) State the modal class.
(b) Draw a histogram to show this information.   AQA

3 The number of words in the first 100 sentences of a book are shown in the table.

Number of words	1 to 10	11 to 20	21 to 30	31 to 40	41 to 50
Frequency	45	38	12	4	1

Draw a frequency polygon for these data.   AQA

4 The table shows the times of arrival of pupils at a village primary school one day.

Time of arrival (t)	Number of pupils
$0830 \leqslant t < 0840$	14
$0840 \leqslant t < 0850$	28
$0850 \leqslant t < 0900$	34
$0900 \leqslant t < 0910$	4

(a) Draw a frequency diagram for the data.
(b) Pupils arriving after 0900 are late. What percentage of pupils were late?

5 The graph shows the time taken to score the first goal in 20 football matches.

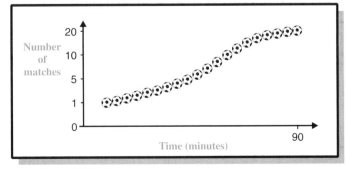

Explain why the graph is misleading.

6 A manager recorded how long each customer spent in his shop.
The table shows his results.

Time (t minutes)	$0 < t \leqslant 10$	$10 < t \leqslant 20$	$20 < t \leqslant 30$	$30 < t \leqslant 40$
Frequency	4	22	18	12

(a) Draw a frequency diagram to represent this data.
(b) Which class interval is the modal class?
(c) As each customer left the shop the manager gave them a questionnaire containing the following question.

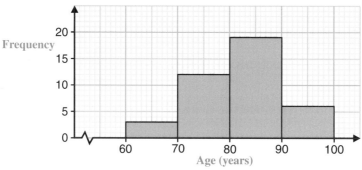

Question:	*How much money did you spend in the shop today?*
Response:	Less than £10 ☐ Less than £20 ☐ Less than £30 ☐ £30 or more ☐

Write down one reason why the response section of this question is not suitable. AQA

7 The graph shows the age distribution of people in a nursing home.

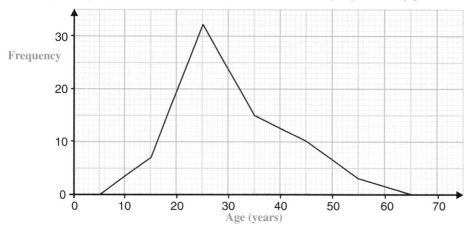

(a) Which age group is the modal class?
(b) How many people are in the nursing home?
(c) The table shows the age distribution of men in the home.

Age (a years)	$60 \leqslant a < 70$	$70 \leqslant a < 80$	$80 \leqslant a < 90$	$90 \leqslant a < 100$
Frequency	2	7	6	0

(i) Draw a frequency polygon to represent this information.
(ii) On the same diagram draw a frequency polygon to represent the age distribution of women in the home.
(iii) Compare and comment on the ages of men and women in the home.

8 The frequency polygon illustrates the age distribution of people taking part in a marathon.

(a) How many people were under 20 years of age?
(b) How many people were over 50 years of age?
(c) How many people took part?

Scatter Graphs ●●●●●●●●●●●●

What you need to know

- A **scatter graph** can be used to show the relationship between two sets of data.
- The relationship between two sets of data is referred to as **correlation**.
- You should be able to recognise **positive** and **negative** correlation. The correlation is stronger as points get closer to a straight line.
- When there is a relationship between two sets of data a **line of best fit** can be drawn on the scatter graph.
- **Perfect correlation** is when all the points lie on a straight line.
- The line of best fit can be used to **estimate** the value from one set of the data when the corresponding value of the other set is known.

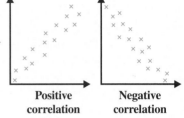

Positive correlation   Negative correlation

Eg 1 The table shows the weights and heights of 10 girls.

Weight (kg)	33	36	37	39	40	42	45	45	48	48
Height (cm)	133	134	137	140	146	146	145	150	152	156

(a) Draw a scatter graph for the data.
(b) What type of correlation is shown?
(c) Draw a line of best fit.
(d) A girl weighs 50 kg. Estimate her height.

> Mark a cross on the graph to show the weight and height of each girl.

(a)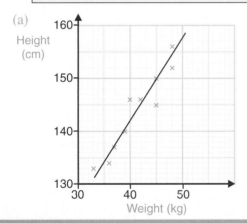

(b) Positive correlation.
(c) The line of best fit has been drawn, by eye, on the graph.

> **On a scatter graph:**
> The **slope** of the line of best fit shows the **trend** of the data.
> The line of best fit does not have to go through the origin of the graph.

(d) 158 cm.
Read estimate where 50 kg meets line of best fit.

Exercise 39

1 The table gives information about the engine size (in cc's) and the fuel economy (in kilometres per litre) of a number of cars.

Engine size (cc)	1800	1000	1200	1600	1400	800	2000	1500
Fuel economy (km/l)	6.5	11	10.5	8	9.5	12	6	8.5

(a) Draw a scatter graph to show this information.
Label the horizontal axis **Engine size (cc)** from 600 to 2000.
Label the vertical axis **Fuel economy (km/l)** from 5 to 15.
(b) Describe the relationship between engine size and fuel economy.
(c) Draw a line of best fit.
(d) Explain how you can tell the relationship is quite strong.

2 The scatter graphs below show the results of a questionnaire given to pupils who have jobs.

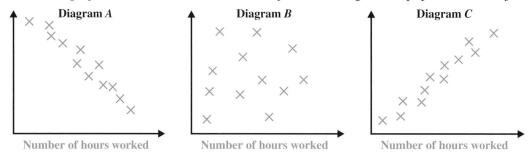

Diagram *A*   Diagram *B*   Diagram *C*

Number of hours worked   Number of hours worked   Number of hours worked

(a) Which scatter graph shows the number of hours worked plotted against:
    (i) the earnings of pupils,
    (ii) the time spent by pupils watching TV,
    (iii) the time taken by pupils to travel to work?

(b) State which one of the graphs shows a negative correlation.

AQA

3 The scatter graph shows the results of candidates in two examinations in the same subject.

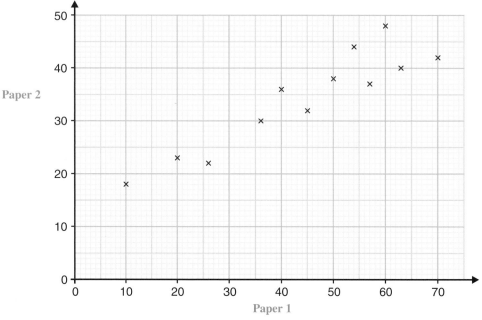

Paper 2

Paper 1

(a) One candidate scored 40 marks on Paper 1.
    What mark did this candidate score on Paper 2?
(b) One candidate scored 48 marks on Paper 2.
    What mark did this candidate score on Paper 1?
(c) Was the highest mark on both papers scored by the same candidate?
(d) Was the lowest mark on both papers scored by the same candidate?
(e) What type of correlation is there between the marks scored on the two exam papers?

4 The table shows the ages and heights of trees in a wood.

Age (years)	1	3	4	5	7	9	10
Height (m)	0.5	1.2	1.7	2.5	3.3	4.5	4.8

(a) Draw a scatter graph for the data.
(b) Draw a line of best fit.
(c) Use the graph to estimate the height of a tree which is
    (i) 8 years old,   (ii) 13 years old.
(d) Which of your answers in (c) is more likely to be reliable?
    Give a reason for your answer.

AQA

Probability

What you need to know

- **Probability** describes how likely or unlikely it is that an event will occur.
  Probabilities can be shown on a probability scale.

  > Probability **must** be written as a **fraction**, a **decimal** or a **percentage**.

Less likely	More likely
Impossible	Certain

  $0 \qquad \frac{1}{2} \qquad 1$

- How to work out probabilities using **equally likely outcomes**.

  > The probability of an event $= \dfrac{\text{Number of outcomes in the event}}{\text{Total number of possible outcomes}}$

  Eg 1 A box contains 7 red pens and 4 blue pens. A pen is taken from the box at random.
  What is the probability that the pen is blue?

  $P(\text{blue}) = \dfrac{\text{Number of blue pens}}{\text{Total number of pens}} = \dfrac{4}{11}$

  > P(blue) stands for the probability that the pen is blue.

- How to estimate probabilities using **relative frequency**.

  > Relative frequency $= \dfrac{\text{Number of times the event happens in an experiment (or in a survey)}}{\text{Total number of trials in the experiment (or observations in the survey)}}$

  Eg 2 A spinner is spun 20 times. The results are shown.

4	1	3	1	4	2	2	4	3	3
4	1	4	4	3	2	2	1	3	2

  What is the relative frequency of getting a 4?

  Relative frequency $= \dfrac{\text{Number of 4's}}{\text{Number of spins}} = \dfrac{6}{20} = 0.3$

  > Relative frequency gives a better estimate of probability the larger the number of trials.

- How to use probabilities to **estimate** the number of times an event occurs in an **experiment** or **observation**.

  > Estimate = total number of trials (or observations) × probability of event

  Eg 3 1000 raffle tickets are sold. Alan buys some tickets.
  The probability that Alan wins first prize is $\frac{1}{50}$.
  How many tickets did Alan buy?   Number of tickets $= 1000 \times \frac{1}{50} = 20$

- **Mutually exclusive events** cannot occur at the same time.

  > When A and B are mutually exclusive events: P(A or B) = P(A) + P(B)

  Eg 4 A box contains red, green, blue and yellow counters.
  The table shows the probability of getting each colour.

Colour	Red	Green	Blue	Yellow
Probability	0.4	0.25	0.25	0.1

  A counter is taken from the box at random.
  What is the probability of getting a red or blue counter?

  P(Red or Blue) = P(Red) + P(Blue) = 0.4 + 0.25 = 0.65

103

The probability of an event, A, **not happening** is: P(not A) = 1 − P(A)

Eg 5 Kathy takes a sweet from a bag at random.
The probability that it is a toffee is 0.3.
What is the probability that it is **not** a toffee?
P(not toffee) = 1 − P(toffee) = 1 − 0.3 = 0.7

● How to find all the possible outcomes when two events are combined.
By **listing** the outcomes systematically.
By using a **possibility space diagram**.

Exercise 40

1 (a) The list gives some words used in probability.

impossible   unlikely   evens   likely   certain

For each of these events, write down the word which describes its probability.
(i)   A fair coin landing on heads.
(ii) Picking a red ball, at random, from a bag containing 20 red balls and 3 black balls.
(iii) Throwing the number 8 on an ordinary, fair, six-sided dice.

(b) Brian has some red marbles, blue marbles and white marbles in a bag.
He says that the probability of choosing each colour is shown in the table.

Colour of marble	red	blue	white
Probability	0.3	0.6	0.2

There is a mistake in the probabilities in the table. Explain how you know this.   AQA

2 An adult is chosen at random in Liverpool.
The probabilities of four events are marked on the probability scale.

    A: The adult is a female.   **C:**   The adult is in Liverpool.
    B: The adult was born in France.   **D:**   The adult is left-handed.

(a) Copy the scale and label each arrow with the correct letter.
(b) Use the scale to estimate the probability that the adult is left-handed.
(c) Estimate the probability that the adult is **not** left-handed.   AQA

3 A packet contains 1 red balloon, 3 white balloons and 4 blue balloons.
A balloon is taken from the packet at random.
What is the probability that it is   (a) red,   (b) red or white,   (c) not white?

4 The diagram shows two sets of cards. One card is taken from each set at random.

Set A   Set B

(a) List all the possible outcomes.
(b) The numbers on the cards are added together to give a score.
What is the probability of getting a score of 6?   AQA

5 The letters of the word   A B B E Y   are written on separate cards and placed in a box.
A card is taken from the box at random.
(a) What is the probability that it is the letter B?
(b) The probability that it is a vowel is 0.4
What is the probability that it is not a vowel?

6 Chris plays a game.
He has two sets of four cards.
The first set of cards is numbered 1 to 4.
The second set of cards is numbered 5 to 8.

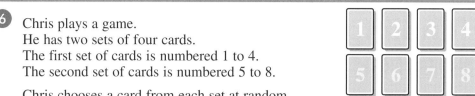

Chris chooses a card from each set at random.
He adds together the numbers on the two chosen cards to get his score.

(a) Copy and complete the table to show all the scores that Chris can make.

1st set of cards

		1	2	3	4
2nd set of cards	**5**				
	6				
	7				
	8				

(b) Calculate the probability that Chris scores (i) 6, (ii) 10 or more.

(c) Chris plays the game 100 times.
How many times would you expect him to score 9?

AQA

7 Petra has 5 numbered cards. She uses the cards to do this experiment:

> Shuffle the cards and then record the number on the top card.

She repeats the experiment 20 times and gets these results.

3	3	2	3	4	3	5	2	3	4
3	5	3	3	4	2	5	3	4	2

(a) What is the relative frequency of getting a 3?

(b) What numbers do you think are on the five cards? Give a reason for your answer.

(c) She repeats the experiment 500 times.
Estimate the number of times she will get a 5. Give a reason for your answer.

8 The table shows information about the colour and type of symbol printed on some cards.

Colour of symbol

		Red	Yellow	Blue
Type of symbol	**O**	9	4	5
	X	2	7	3

(a) A card is taken at random.
   (i) What is the probability that it has a red symbol?
   (ii) What is the probability that it has a blue symbol **or** an X?

(b) A yellow card is taken at random. What is the probability that it has the symbol X? AQA

9 A box contains counters. The counters are numbered 1, 2, 3, 4 or 5.
A counter is taken from the box at random.

(a) Copy and complete the table to show the probability of each number being chosen.

Number on counter	1	2	3	4	5
Probability	0.20	0.30	0.15		0.10

(b) Is the number on the counter chosen more likely to be odd or even?
You must show your working.

AQA

10 In a sixth form college there are 1000 students. $\frac{2}{5}$ of the students are girls.

The probability that a girl chosen at random studies mathematics is $\frac{1}{10}$.

The probability that a boy chosen at random studies mathematics is $\frac{1}{6}$.
Calculate how many students in the college study mathematics.

AQA

Handling Data
Non-calculator Paper

Do not use a calculator for this exercise.

1 The pictogram shows the number of videos hired from a shop each day last week.

Monday	⊕⊕ ⊕
Tuesday	⊕⊕ ⊕⊕
Wednesday	⊕⊕ ◠
Thursday	⊕⊕ ⊕⊕ ◖⊙
Friday	⊕⊕ ⊕⊕ ⊕⊕ ⊕⊕ ⊕
Saturday	

On Monday 6 videos were hired.

(a) How many videos does ⊕⊕ represent?

(b) How many videos were hired on Thursday?

70 videos were hired altogether last week.

(c) How many videos were hired on Saturday?

2 Paige did a survey about pets. She asked each person,

"How many pets do you have?"

Here are her results.

    3 2 1 1 4 2 3 1 0 0 1 3 0

    1 4 2 0 1 5 1 2 4 1 4 2

(a) Copy and complete the frequency table for this data.

Number of pets	Tally	Frequency
0		
1		

(b) Draw a bar chart to show this data.

3 The results of a survey of the holiday destinations of people booking holidays abroad are shown in the bar chart.

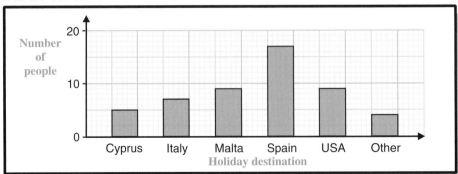

(a) Which holiday destination is the mode?
(b) How many more people are going to Spain than to Cyprus?
(c) How many people are included in the survey?

4 Vikram records the number of tennis matches won by each of ten players
His results are:

$$8, \quad 7, \quad 9, \quad 9, \quad 4, \quad 2, \quad 8, \quad 2, \quad 3, \quad 8.$$

(a) Write down the mode.
(b) Work out the median.
(c) Calculate the mean.
(d) Find the range.

AQA

5 The graph shows the distribution of the best height jumped by each girl in a high jump competition.

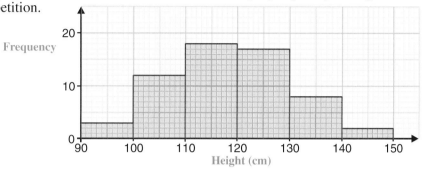

(a) How many girls jumped less than 100 cm?
(b) How many girls jumped between 100 cm and 120 cm?
(c) How many girls took part in the competition?

6 Sylvester did a survey to find the most popular pantomime.

(a) The results for children are shown in the table.

Pantomime	Aladdin	Cinderella	Jack and the Bean Stalk	Peter Pan
Number of children	45	35	25	15

(i) Draw a clearly labelled pie chart to illustrate this information.
(ii) Which pantomime is the mode?

(b) The results for adults are shown in the pie chart.

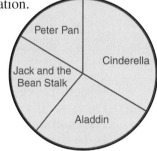

(i) 20 adults chose Aladdin.
How many adults were included in the survey?
(ii) What percentage of adults chose Cinderella?

7

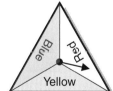

Karl plays a game with a spinner.
The spinner has three equal sections, coloured red, yellow and blue.
Karl spins the spinner twice.
If both spins land on the same colour, Karl wins 2 tokens.
If exactly one of the spins lands on red, Karl wins 1 token.
For any other result, Karl wins 0 tokens.

(a) Copy and complete the table to show the numbers of tokens that Karl can win.

		Second spin		
		Red	Yellow	Blue
	Red			
First spin	Yellow			
	Blue			

(b) What is the probability that Karl wins 0 tokens?
(c) Karl plays the game 70 times.
How many times should he expect to win 2 tokens?

AQA

8 Linzi is doing a survey to find if there should be a supermarket in her neighbourhood. This is one of her questions.

> "Do you agree that having a supermarket in the neighbourhood would make it easier for you to do your shopping and if we did have one would you use it?"

Give two reasons why this question is unsuitable in its present form.

9 Winston has designed a data collection sheet to record the number of bottles that each person puts into a bottle bank.

Number of bottles	0 to 2	3 to 6	6 to 8
Tally			
Frequency			

(a) Give **three** criticisms of the class intervals that Winston has chosen.

Anna and Patrick watch people using the bottle bank.
Anna watches 60 people and calculates the mean to be 8.5 bottles per person.
Patrick watches 15 people and calculates the mean to be 9.2 bottles per person.

(b) Which of the two means would you expect to give the more reliable estimate of the mean number of bottles per person? Give a reason for your answer.

AQA

10 The lengths of 20 bolts, in centimetres, is shown.

    7.4   5.8   4.5   5.0   6.5   6.6   7.0   5.4   4.8   6.4
    5.4   6.2   7.2   5.5   4.8   6.5   5.0   6.0   6.5   6.8

(a) Draw a stem and leaf diagram to show this information.
(b) What is the range in the lengths of these bolts?

11 The table shows information about a group of students.

	Can speak French	Cannot speak French
Male	5	20
Female	12	38

(a) One of these students is chosen at random.
What is the probability that the student can speak French?
(b) Pru says, "If a female student is chosen at random she is more likely to be able to speak French than if a male student is chosen at random."
Is she correct? Explain your answer.

12 The table shows the rainfall (cm) and the number of hours of sunshine for various towns in August one year.

Rainfall (cm)	0.1	0.1	0.2	0.5	0.8	1	1	1.5	1.5	1.9
Sunshine (hours)	200	240	210	190	170	160	130	100	120	90

(a) Use this information to draw a scatter graph.
(b) Draw a line of best fit on your diagram.
(c) Use your line of best fit to estimate the number of hours of sunshine for a town that had:
    (i) 1.4 cm of rain in August that year,
    (ii) 2.5 cm of rain in August that year.
(d) Which of the answers from part (c) would you expect to be the more reliable?
Give a reason for your answer.

AQA

13 A bag contains 50 cubes of which 7 are red. A cube is taken from the bag at random.
(a) The probability that it is white is 0.3.
What is the probability that it is not white?
(b) What is the probability that it is either white or red?

Handling Data
Calculator Paper

You may use a calculator for this exercise.

1 The table shows the number of books borrowed from a library during five days.

Day	Monday	Tuesday	Wednesday	Thursday	Friday
Number of books	40	35	30	15	50

(a) How many books were borrowed during these five days?

(b) Draw a pictogram to represent the information. Use ▭ to represent 10 books.

2 Five children guess the score on the next throw of a fair, six-sided dice.

Abdul: It will be an even number. Barbara: It will be a five.
David: It will be a seven. Cathy: It will be less than five.
Ewan: It will be a number less than seven.

The scale shows the probability of each statement being correct.

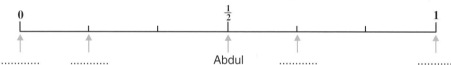

Copy the scale and fill in the names of the children. Abdul's has been done for you. AQA

3 Nine people were asked to estimate the height, in metres, of a building.
The estimates the people made are shown.

          7   12   10   9   11   12   10   12   11

(a) Which height is the mode?
(b) Work out the median height.
(c) Calculate the mean height. AQA

4 Pat carried out a survey.
She asked each pupil in her class how many postcards they received last August.
Her results are shown in the vertical line graph.

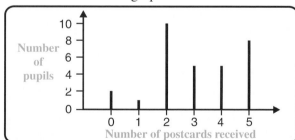

(a) What is the modal number of postcards received?
(b) How many pupils took part in the survey?
(c) How many postcards were received altogether? AQA

5 Karina is playing a game with these cards. [X] [Y] [1] [1] [3]

One card is taken at random from the letters.
One card is taken at random from the numbers.
(a) List all the possible outcomes.
(b) Explain why the probability of getting [X] [1] is not $\frac{1}{4}$.

6 The table shows the number of peas in a sample of pods.

Number of peas	1	2	3	4	5	6	7	8
Number of pods	0	0	2	3	5	7	2	1

(a) How many pods were in the sample?
(b) What is the modal number of peas in a pod?
(c) What is the range in the number of peas in a pod?
(d) Draw a bar chart to show this information.

7 The stem and leaf diagram shows the weights, in grams, of letters posted by a secretary.

(a) How many letters were posted?

(b) What is the median weight of one of these letters?

(c) What is the range in the weights of these letters?

(d) Calculate the mean weight of a letter?

```
                              1 | 5  means 15 grams
              1 | 5   8
              2 | 0   4   5   6   8   8
              3 | 1   2   3   5   7
              4 | 2   5
```

8 The bar chart shows the amount of time Year 6 pupils spend doing homework and watching television.

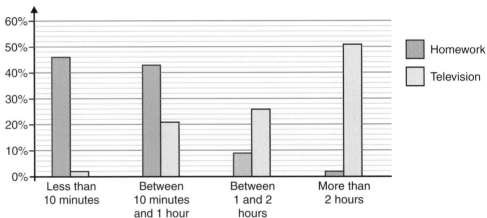

(a) What percentage of these pupils spend more than 2 hours watching television?
(b) Use the bar chart to complete this table for 'Time doing homework'.

Time doing homework	Less than 10 minutes	Between 10 minutes and 1 hour	Between 1 and 2 hours	More than 2 hours
Percentage of Year 6 pupils	46%			

(c) In a survey, Year 11 pupils were asked the question:

"Where do you learn the most?"

Their replies are shown in the pie chart.
(i) What was the most common reply to the question?
(ii) What percentage of pupils said they learnt most from television?
(iii) What fraction of the pupils said they learnt most at home?

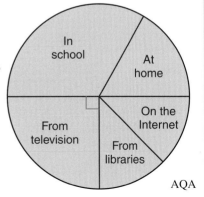

AQA

9 The mean weight of 7 netball players is 51.4 kg.
(a) Find the total weight of the players.

The mean weight of the 7 players and the reserve is 52.3 kg.
(b) Calculate the weight of the reserve.

AQA

10 A baby elephant is weighed at the end of each week. The table shows the weights recorded.

Week	1	2	3	4	5	6	......	10
Weight (kg)	85	92	94	98	100	104	......	120

(a) Plot these points on a graph, and join with straight lines.
(b) Use your graph to estimate the weight of the elephant at the end of week 8.

AQA

11 The table shows the results of asking 480 people how they travel to work.

Method of travel	Bus	Train	Car	Walk
Number of people	120	80	180	100

Draw a clearly labelled pie chart to represent this information.

AQA

12 Mary has a bag in which there are 8 marbles, all green.
Tony has a bag in which there are 12 marbles, all red.
Jane has a bag in which there are some blue marbles.

(a) What is the probability of picking a red marble from Mary's bag?
(b) Mary and Tony put all their marbles into a box.
What is the probability of choosing a red marble from the box?
(c) Jane now adds her blue marbles to the box.
The probability of choosing a blue marble from the box is now $\frac{1}{2}$.
How many blue marbles does Jane put in the box?

AQA

13 Corrin throws a dice 40 times. Her results are shown.

Score	1	2	3	4	5	6
Frequency	7	6	7	6	6	8

(a) Which score is the mode?
(b) Calculate the mean score.
(c) What is the median score?

14 The numbers of people exposed to different types of radiation in the UK were recorded.
The pie chart shows the results.

If 12 000 people were exposed to Gamma radiation last year, estimate the total number of people who were exposed to any form of radiation last year.

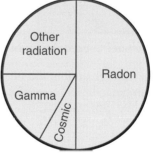

AQA

15 The table shows the weight distribution of the fish caught in a fishing competition.

Weight (g grams)	Frequency
$0 \leqslant g < 100$	0
$100 \leqslant g < 200$	16
$200 \leqslant g < 300$	36
$300 \leqslant g < 400$	20
$400 \leqslant g < 500$	8
$500 \leqslant g < 600$	0

(a) Calculate an estimate of the mean weight of a fish.
(b) Draw a frequency polygon to represent this distribution.

AQA

Do not use a calculator for this exercise.

1   (a)   Write the number   two thousand six hundred and nine   in figures.
      (b)   Write the number   60 000 000   in words.

2   The table shows the number of cars parked in an office car park each day.

Day	Monday	Tuesday	Wednesday	Thursday	Friday
Number of cars	8	12	6	9	10

   (a)   How many more cars were parked on Tuesday than on Thursday?
   (b)   Draw a pictogram to represent the information.   Use ⊕ to represent 4 cars.

3   (a)   Put these numbers in order of size, smallest first.

   | 105 | 30 | 7 | 19 | 2002 |

   (b)   Work out.   (i) $105 - 30$ (ii) 19×7 (iii) $2002 \div 7$

4   A football stadium has 57 896 seats.
   Write the number   57 896   to the nearest thousand.   AQA

5   Bananas cost 89p per kilogram.
   (a)   (i)   How much do 3 kg of bananas cost?
         (ii)   Jane buys 3 kg of bananas and pays with a £5 note.
               How much change should she get?

BANANAS
89p
per kilogram

   (b)      Barry also bought some bananas.
               The weight of bananas he bought is shown on the scales.
               What weight of bananas did Barry buy?

   (c)   Michael bought 3.5 kg of bananas. How many grams is this?   AQA

6   Write a rule for finding the next number in each sequence and use your rule to find the
   next number.
   (a)   3,   9,   15,   21,   27,   …
   (b)   1,   2,   4,   8,   16,   …

7   The diagram shows a rectangle and a triangle
   drawn on 1 cm squared paper.
   (a)   How many lines of symmetry has
         (i)   the rectangle,
         (ii)   the triangle?
   (b)   What is the perimeter of the rectangle?
   (c)   What is the area of the triangle?

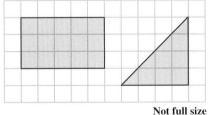

Not full size

8   (a)   A roll costs d pence. How much will 5 rolls cost?
   (b)   A cake costs 25 pence more than a roll. How much does a cake cost?

9   There are 160 people at a funfair and $\frac{1}{4}$ of them are wearing shorts.
   How many are wearing shorts?   AQA

10   (a)   Write down the smallest number which is a multiple of both 3 and 5.
    (b)   Write down all the factors of 6.

11 Clare makes a solid shape by joining together cubes of side 1 cm.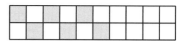
    (a) Work out the volume of the solid.
    (b) Before she made the solid, she had 30 cubes.
        What fraction of the cubes did she use?
                                   AQA

12 Use the formula $P = 5m + 2n$ to find the value of P when $m = 4$ and $n = 3$.

13 (a) What fraction of the rectangle is shaded?
        Give your answer in its simplest form.
    (b) What percentage of the rectangle is **not** shaded?

14 Sports Wear Ltd. hire out ski-suits. The cost of hiring a ski-suit is calculated using this rule.

> Four pounds per day plus a fixed charge of five pounds.

    (a) How much would it cost to hire a ski-suit for 8 days?
    (b) Heather paid £65 to hire a ski-suit.
        For how many days did she hire it?

15 (a) On graph paper plot the points $P(4, 1)$ and $Q(2, -5)$.
    (b) Find the coordinates of the midpoint of the line segment PQ.

16 The prices, in pence, of 9 different loaves of bread are shown.

$$23, \quad 36, \quad 69, \quad 49, \quad 38, \quad 55, \quad 82, \quad 69, \quad 29$$

Work out   (a) the range in prices,   (b) the median price,   (c) the mean price.

17 The temperature at noon was 5°C. At midnight the temperature was 7°C colder.
What was the temperature at midnight?

18 In the diagram, XY is a straight line.
Find the size of the angle a.

19 The diagram shows part of a map of Tasmania.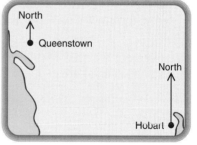
    (a) What is the three-figure bearing of
        Queenstown from Hobart?
    (b) The actual distance between Queenstown
        and Hobart is 162 km.
        Calculate the distance between Queenstown
        and Hobart in miles.
                                 AQA

20 (a) Work out (i) $7 - 3.72$,   (ii) $\frac{3}{5}$ of 9,   (iii) $3 + 4 \times 8$.
    (b) What is the value of $5^2 + \sqrt{36}$?

21 (a) Simplify. (i) $a + 2a + 4a$ (ii) $2x + 5y - 3x + 6y$ (iii) $a \times a \times 3$
    (b) Solve.   (i) $5x = 15$ (ii) $3x - 2 = 10$ (iii) $2x + 1 = 7$
    (c) Find the value of $3x + y^3$ when $x = -1$ and $y = 2$.

22 Each term in a sequence of numbers is obtained by multiplying the previous terms by -2.
The first three terms are: 1, -2, 4, …
    (a) Write down the next two terms in the sequence.
    (b) Will the 50th term of the sequence be positive or negative?
        Explain your reasoning.
                                      AQA

23 (a) Draw and label the lines $y = x + 1$ and $x + y = 3$ for values of x from -1 to 3.
    (b) Write down the coordinates of the point where the lines cross.

24 Susan and Jill play a game.

   (a) Susan has a box containing 3 red, 4 yellow and 2 blue counters.
       She picks a counter at random.
       What is the probability that Susan picks a yellow counter?

   (b) Jill has a box containing 18 counters, of which 8 are yellow.
       She picks a counter at random.
       What is the probability that Jill does **not** pick a yellow counter?

   (c) Who is more likely to pick a yellow counter?   AQA

25 To calculate the number of mince pies, m, to make for a Christmas Party for p people,
Donna uses the formula $m = 2p + 10$.

   (a) How many mince pies would she make for a party of 12 people?

   (b) Donna makes 60 mince pies for another party.
       How many people are expected at this party?

26 (a) In the diagram, angle $BCD = 76°$, $AC = BC$
       and ACD is a straight line.
       Work out the size of angle BAC.

   (b) Information about some triangles is shown.

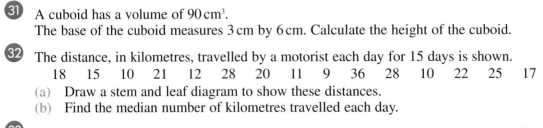

       Which two of these triangles are congruent to each other? Give a reason for your answer.

27 (a) A postwoman has only 1st class and 2nd class letters in her post bag.
       80% of the letters are 1st class. There are 320 letters altogether.
       How many letters are 1st class?

   (b) A mail van has 9000 letters and 150 parcels.
       Express the number of letters to the number of parcels as a ratio in its simplest form.
                                                       AQA

28 (a) Multiply out $5(x + 3)$.

   (b) Solve the equations   (i) $2(2x + 5) = 34$,   (ii) $5x = 6 + x$.

29 Three-tenths of the area of this shape is shaded.
Calculate the shaded area.

                                                           AQA

30 The diagram shows a sketch of a regular pentagon.
       Work out the size of angle x.

31 A cuboid has a volume of $90\,cm^3$.
The base of the cuboid measures $3\,cm$ by $6\,cm$. Calculate the height of the cuboid.

32 The distance, in kilometres, travelled by a motorist each day for 15 days is shown.

   18   15   10   21   12   28   20   11   9   36   28   10   22   25   17

   (a) Draw a stem and leaf diagram to show these distances.

   (b) Find the median number of kilometres travelled each day.

33 The numbers on these cards are coded. The sum of the numbers on these 3 cards is 41.

   (a) Form an equation in x. | x | | $2x - 1$ | | $3x$ |

   (b) By solving your equation, find the numbers on the cards.

34 Work out   (a) $\frac{2}{5} \times \frac{1}{4}$ (b) 0.4×0.2 (c) $2^3 \times 5^2$ (d) $4\frac{2}{3} + 1\frac{3}{5}$ AQA

35 Write 72 as a product of its prime factors.

36 A farmer has two crop circles in his field.
One circle has a radius of 9 m and the other has a diameter of 12 m.
(a) What is the ratio of the diameter of the small circle to the diameter of the large circle? Give your answer in its simplest form.
(b) Calculate, in terms of π, the circumference of the smaller circle.

37 A concrete block weighs 11 kg, correct to the nearest kilogram.
Write down the greatest and least possible weight of the block.

38 Sheila lives 6 kilometres from the beach.
She jogs from her home to the beach at an average speed of 10 km/h.
She gets to the beach at 1000. Calculate the time when she left home.   AQA

39 A sequence begins: 2, 5, 8, 11, …
Write in terms of n, the nth term of the sequence.

40 The diagram shows the positions of shapes P, Q and R.

(a) Describe fully the single transformation which takes P onto Q.

(b) Describe fully the single transformation which takes P onto R.

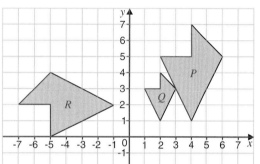

41 Twenty pupils each shuffle a pack of coloured cards and choose a card at random.
The colour of the card is recorded for each pupil.

(R = Red   B = Blue   G = Green   Y = Yellow)

B   G   Y   B   Y   R   R   B   Y   Y
B   G   G   B   B   R   R   B   Y   Y

(a) Use these results to calculate the relative frequency of each colour.
(b) Use the results to calculate how many times you would expect a blue card if 100 pupils each choose a card at random.   AQA

42 (a) Copy and complete the table of values for $y = x^2 - 3x + 1$.

x	-1	0	1	2	3	4
y		1	-1			5

(b) Draw the graph of $y = x^2 - 3x + 1$ for values of x from -1 to 4.
(c) Use your graph to find the value of y when $x = 1.5$.
(d) Use your graph to solve the equation $x^2 - 3x + 1 = 0$.

43 Cocoa is sold in cylindrical tins. The height of a tin is 7.9 cm. The radius of a tin is 4.1 cm.
Use approximations to estimate the volume of a tin. Show all your working.

44 (a) Factorise. (i) $5a - 10$ (ii) $x^2 - 6x$
(b) Expand and simplify. (i) $3(a + 4) - 5(2 - a)$ (ii) $(x - 2)(x - 4)$
(c) Make t the subject of the formula. $W = 5t + 3$
(d) Simplify. $m^2 \times m^3$

45 These formulae represent quantities connected with containers, where a, b and c are dimensions. $2(ab + bc + cd)$ abc $\sqrt{a^2 + b^2}$ $4(a + b + c)$
Which of these formulae represent lengths? Explain how you know.

46 Solve the inequality $3 < 2x + 1 < 5$.   AQA

Calculator Paper

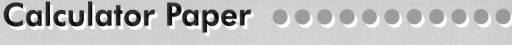

You may use a calculator for this exercise.

1 (a) Write these numbers in order, smallest first:
$$5, \quad -7, \quad 0, \quad 3.5, \quad 10$$

(b) What is the value of the 8 in the number 18 326?

2 The diagram shows four patterns in a sequence.

```
                                        ＊ ＊
                         ＊ ＊          ＊ ＊ ＊
           ＊ ＊        ＊ ＊ ＊        ＊ ＊ ＊ ＊
  ＊ ＊    ＊ ＊ ＊      ＊ ＊ ＊ ＊      ＊ ＊ ＊ ＊ ＊
Pattern 1  Pattern 2     Pattern 3         Pattern 4
```

(a) Draw the next pattern in the sequence.

(b) Copy and complete the table.

Pattern number	1	2	3	4	5
Number of stars	2	5	9	14	

(c) How many stars will be in Pattern 6?

3 A series of measurements is given.

120 centimetres, 12 feet, 400 metres, 200 miles, 2000 kilometres

Which of these measurements is a sensible estimate of:
(a) the length of a car;
(b) the perimeter of a running track;
(c) the distance between London and Paris?

AQA

4 Copy the diagram and draw a reflection of the shape in the mirror line *PQ*.

5 These words are used in probability.

impossible   unlikely   evens   likely   certain

A fair, six-sided dice is thrown once.
Which word describes the probability of getting each of these events?
(a) An odd number.   (b) A number **more** than 5.

6 (a) From the list below, choose the correct word to describe triangle *T*.
isosceles   equilateral   right-angled   scalene

(b) How many sides does a hexagon have?

(c)
(i) Choose the correct word to describe quadrilateral *Q*.
square   kite   rectangle   trapezium
(ii) Copy the diagram and draw 3 more quadrilaterals, like *Q*, to show how this shape tessellates.   AQA

7 The diagram shows two gear wheels.
The large wheel has 24 teeth.
The small wheel has 12 teeth.
Describe what happens to the small wheel
when the large wheel is turned through 90° in a clockwise direction.

8 In France a bicycle costs 120 euros. £1 = 1.40 euros.
How much is the bicycle in £s, correct to the nearest £?

9 These sections are from rulers that have had some of their markings rubbed off. What are the values of *a* and *b* on the rulers?

(a) 7   9   *a*   (b) 0.6   1.0   *b*

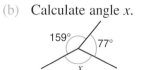

AQA

10 Ali is twice as old as Sue. Sue is 2 years younger than Philip. Philip is 11 years old.
(a) How old is Sue?   (b) How old is Ali?   AQA

11 (a)
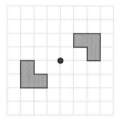

Triangle *ABC* is isosceles, with *AB* = *AC* and angle *B* = 56° Calculate the size of angle *A*.

(b) Calculate angle *x*.

159° 77°

x

AQA

12 Copy the diagram and draw two more shapes so that the final pattern has rotational symmetry of order 4.

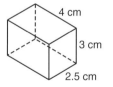

13 Write $\frac{1}{3}$, 0.5, 40% in order, smallest first.

14 The cost of printing wedding invitation cards is calculated using this formula.

| 30p per card plus £7.50 |

(a) Mandy has 50 cards printed. How much does it cost?
(b) It costs Bob £31.50 to have some cards printed. How many cards did he have printed?

15 (a) Simplify $5g - 3g + 2g$.
(b) Solve the equations (i) $3n = 12$,   (ii) $3m + 1 = 10$.
(c) Find the value of $2h^2$ when $h = 3$.
(d) Find the value of $3p + q$ when $p = -2$ and $q = 5$.

16 The diagram shows a cuboid.
(a) Draw an accurate net of the cuboid.
(b) Work out the area of the net.

4 cm
3 cm
2.5 cm

17 An input-output diagram is shown.

Input → -5 → $\times 4$ → Output

(a) When the INPUT is 8, what is the OUTPUT?
(b) When the OUTPUT is 36, what is the INPUT?   AQA

18 The pie chart shows the proportions of complaints made about different parts of the Health Service last year.

(a) What fraction of complaints were made about doctors?
(b) There were 400 complaints made about hospitals. How many complaints were made altogether?
(c) Work out the number of complaints made about dentists.

AQA

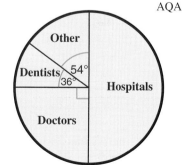

Other

Dentists 54°
36°   Hospitals

Doctors

19 Calculate. (a) $\sqrt{9.61}$ (b) $\sqrt{9.61} + 2.9^2$ AQA

20 The first four numbers in a sequence are: 15, 11, 7, 3, ...
(a) What are the next two numbers in the sequence?
(b) Explain how you found your answers.

21 (a) Simplify $5m - m + 5$.
   (b) Multiply out $3(t - 5)$.
   (c) Solve (i) $3x = 1$,   (ii) $3 = 19 + 4y$.
   (d) Find the value of $m^3 - 3m$ when $m = -2$.

22 (a) The diagram shows a regular pentagon.
      How many lines of symmetry does a regular pentagon have?

   (b)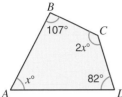

   ABCD is a quadrilateral.
   Work out the value of *x*.

AQA

23 (a) Draw the graph of $y = 2x - 3$ for values of *x* from -2 to $+3$.
   (b) The line $y = 2$ crosses $y = 2x - 3$ at *P*.
      Write down the coordinates of *P*.

AQA

24 A fairground ride is decorated with 240 coloured lights.
   (a) 15% of the lights are red. How many red lights are there?
   (b) 30 of the 240 lights are not working. What percentage of lights are not working?

25 (a) *k* is an even number. Jo says that $\frac{1}{2}k + 1$ is always even.
      Give an example to show that Jo is wrong.
   (b) The letters *a* and *b* represent prime numbers.
      Give an example to show that $a + b$ is **not** always an even number.

AQA

26 A jigsaw puzzle is a rectangle measuring 17.6 cm by 8.5 cm.
   35% of the area of the puzzle is blue.
   Calculate the area of the puzzle which is blue.

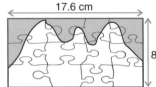

AQA

27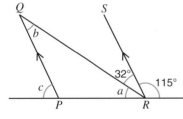

   In the diagram, *PQ* is parallel to *RS*.
   Find the size of angles *a*, *b* and *c*.

28 A hospital carries out a test to compare the reaction times of patients of different ages.
   The results are shown.

Age in years	17	21	24	25	31	15	18	29	20	26
Time (hundredths of a second)	29	40	45	65	66	21	33	62	32	53

   (a) Plot the results as a scatter graph.
   (b) What does the scatter graph tell you about the reaction times of these patients?
   (c) Draw a line of best fit on the scatter graph.
   (d) The hospital is worried about the reaction time of one patient.
      How old is the patient? Give a reason for your answer.

AQA

29 A toilet roll has 240 single sheets per roll.
   Each sheet is 139 millimetres long and 110 millimetres wide.

   Calculate the total area of paper on the roll.
   Give your answer in square metres.

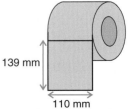

139 mm

110 mm

AQA

30 A hang glider flies 2.8 km on a bearing of 070° from P to Q and then 2 km on a bearing of 200° from Q to R.
   (a) Make a scale drawing to show the flight of the hang glider from P to Q to R.
       Use a scale of 1 cm to 200 m.
   (b) From R the hang glider flies directly back to P.
       Use your drawing to find the distance and bearing of P from R.

31 Solve the equations   (a) $3a - 2 = -5$,   (b) $5m + 3 = 7 - m$,   (c) $3(a - 2) = 6$.

32 A recipe for plum jam states:

   2 kg of plums and 2 kg of sugar are needed to make 3.6 kg of jam.

   (a) Kevin makes 4.5 kg of jam.
       How many kilograms of plums did he need?
   (b) Kevin puts the jam into 1 lb jars.
       How many jars are needed for 4.5 kg of jam?   AQA

33 A solid plastic cuboid has dimensions 3 cm by 5 cm by 9 cm.
   The density of the plastic is 0.95 grams per cm³.
   What is the weight of the plastic cuboid?

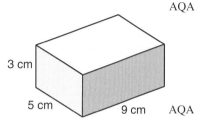

3 cm

5 cm   9 cm   AQA

34 Charles drove 132 miles at an average speed of 55 mph.
   Calculate the time taken for this journey.
   Give your answer in hours and minutes.   AQA

35 Some students took part in a sponsored silence.
   The frequency diagram shows the distribution of their times.

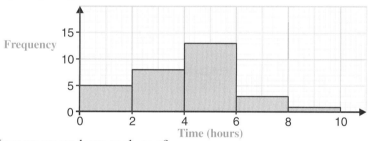

   (a) How many students took part?
   (b) Which time interval contains the median of their times?
   (c) Calculate an estimate of the mean of their times.

36 Use a trial and improvement method to find a solution to the equation $x^3 + x = 57$.
   Show all your working and give your answer correct to one decimal place.

37 The diagram shows a semi-circle with diameter AB.
   C is a point on the circumference.
   $\angle ACB = 90°$.
   $AC = 6$ cm   and $CB = 8$ cm.

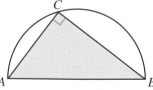

   Calculate the area of the shaded triangle as a percentage of the area of the semi-circle.

38 Calculate $\dfrac{35.6^2}{23.8 \times 22.6}$.
   Give your answer to 3 significant figures.   AQA

39 (a) Write down the values of n, where n is an integer, which satisfies the inequality
$$-1 < n + 2 \leqslant 3.$$
   (b) Solve the inequality $2x + 3 < 4$.

Answers

Exercise 1   Page 1

1. (a) 5 000 000   (b) 605 230
2. 117, 100, 85, 23, 9
3. (a) 3 thousands
   (b) Twenty-three thousand five hundred and forty-seven
4. (a) 2 and 58   (b) 29 and 39
5. (a) 81   (b) (i) 35 (ii) 100 (iii) 10
6. (a) 1005   (b) 191   (c) 183
7. (a) (i) 2358   (ii) 8523
   (b) 500   (c) 50   (d) 6165
8. (a) 466 km   (b) Jean's journey by 34 km
9. (a) 18   (b) 6   (c) 3
10. (a) (i) 2   (ii) 8
    (b) E.g. $20 \div 7$ has a remainder of 6.
11. (a) 12 000   (b) 500   (c) 175   (d) 15
12. (a) £2250   16. £35
    (b) £540   17. £181 per month
    (c) £1710   18. 150 cm
13. (a) 352   19. 115 cm
    (b) 12 636   20. 13 301
    (c) 21   21. (a) 3290
14. 50 boxes   (b) 7 reams
15. (a) 18   22. (a) 770
    (b) 12   (b) (i) 19 coaches needed
    (c) 3   (ii) 9 empty seats

Exercise 2   Page 3

1. 0.065 and 0.9
2. 1.08, 1.118, 1.18, 1.80
3. (a) 18.59   (b) 11.37   (c) 2.9   (d) 2.64
4. (a) 10.85 kg   (b) 26.55 kg   (c) 105 dollars
5. £8.65
6. (a) 320   (b) 0.32
7. 74 pence
8. (a) $900 \times 0.6 = 540$
       E.g. $900 \times 0.6 = 9 \times (100 \times 0.6)$
                     $= 9 \times 60$
                     $= 540$
   (b) $40 \div 0.8 = 50$
       E.g. $40 \div 0.8 = 400 \div 8$
                     $= 50$
9. 1.57 m
10. (a) There are two figures after the decimal points in the question but only one in the answer.
    (b) (i) 0.12   (ii) 0.06

11. (a) (i) 4.02   (ii) 12
    (b) (i) 136   (ii) 0.245
12. (a) 22.75   (b) 0.65   (c) 6500
13. (a) $\frac{3}{10}$ (b) $\frac{3}{100}$ (c) $\frac{33}{100}$
14. 4 packets   16. 62 pence
15. 98 pence per kilogram   17. 17.76792453

Exercise 3   Page 6

1. (a) 626   (b) 630   (c) 600
2. (a) 36 700   (b) 37 000
3. 19 500
4. (a) $\frac{28 + 20}{4}$ (b) 12
5. $300 \text{ km} + 100 \text{ km} = 400 \text{ km}$
6. (a) 9000,   1300,   1700
   (b) $\frac{9000}{1300 + 1700} = \frac{9000}{3000} = 3$
7. $\frac{4000}{8} = 500$ cars
8. $50\text{p} \times 800 = £400$
9. (a) $40 \times 20 = 800$
   (b) (i) $2000 \div 40$ (ii) 50
10. (a) 100 is bigger than 97, **and**
         50 is bigger than 49.
    (b) Smaller.   1000 is smaller than 1067, **and**
                    50 is bigger than 48.
11. (a) 6250   (b) 6349
12. 49.5 m
13. (a) 70×30
    (b) $70 \times 30\text{p} = £21$
         So, Isobella's calculation is not correct.
14. $\frac{400 + 200}{40} = \frac{600}{40} = 15$ Answer is wrong.
15. 17 boxes
16. (a) $\frac{9000}{10} \times 90\text{p} = £810$
    (b) 9000 is larger than 8873,
         10 is smaller than 11, and
         90 is larger than 89.9.
17. 400.5 m
18. $\frac{400 \times 3}{0.6} = \frac{1200}{0.6} = 2000$
19. (a) £8.49   (b) £7.50
20. (a) 14.95   (b) 15.0
21. (a) 680   (b) 700
22. No.   For example, an answer of 0.01634…
         is 0.02 to 2 d.p. and 0.016 to 2 s.f.
         0.016 is more accurate.
23. (a) 18.0952381   (b) 18.1
24. 9.2
25. (a) 31.28, correct to 2 d.p.
    (b) $\frac{90 \times 10}{20 + 10} = \frac{900}{30} = 30$
         So, answer to (a) is about right.

1. (a) $-3°C$ (b) $-13°C$
2. (a) Oslo (b) Warsaw
3. 140 m
4. $-9, -3, 0, 5, 7, 17$
5. (a) 5 (b) -15 (c) -50 (d) -2
6. (a) 6 (b) 5 (c) -3 (d) -3
7. $5°C$
8. 29 degrees
9. £191
10. (a) 6 degrees (b) Between 1200 and 1800
11. (a) 13 degrees (b) $6°C$
12. 7 degrees
13. (a) (i) -20 (ii) 21 (b) (i) -4 (ii) 3
14. $20°F$
15. (a) -20 (b) -15
16. (a)

2	-3	4
3	1	-1
-2	5	0

(b) $(-12) + (18) + (-24) = 18 - 36 = -18$

SECTION 5

Exercise **5**
Page 10

1. (a) $\frac{2}{5}$ (b) Shade 4 parts.
2. $\frac{4}{12}$ and $\frac{5}{15}$
3. (a) $\frac{7}{10}$ as $\frac{4}{5} = \frac{8}{10}$ (b) E.g. $\frac{5}{12}$
4. $\frac{19}{40}$
   Change to equivalent fractions with common denominator 120.
   $\frac{4}{10} = \frac{48}{120}, \frac{9}{20} = \frac{54}{120}, \frac{14}{30} = \frac{56}{120}, \frac{19}{40} = \frac{57}{120}$
5. $\frac{7}{12}, \frac{5}{8}, \frac{2}{3}, \frac{3}{4}$
6. £1.80
7. 15 miles
8. 5 tins
9. $\frac{2}{3}$
10. £6650
11. £3.64
12. (a) 0.167 (b) 1.7, 1.67, 1.66, $1\frac{1}{6}$, 1.067
13. (a) $\frac{1}{6}$ (b) $\frac{5}{12}$
14. $\frac{12}{25}$
15. $\frac{9}{20}$
16. (a) $5\frac{9}{10}$ (b) $\frac{7}{15}$ (c) $1\frac{1}{2}$
17. $\frac{5}{6}$ of a cup

Exercise **6**
Page 13

1. (a) 2, 5 (b) 20 (c) 2, 5, 11, 17
2. (a) 1, 2, 3, 6, 9, 18 (b) 35
   (c) 9 has more than 2 factors: 1, 3, 9
3. (a) 36 (b) 10 (c) 27 (d) 2
4. (a) 16 (b) 125 (c) 10 000 (d) 1
5. E.g. $4^2 = 16$
6. No. $2^2 + 3^2 = 4 + 9 = 13$
   $(2 + 3)^2 = 5^2 = 25$
7. (a) 2 (b) 64
8. (a) (i) 125 (ii) 8 (b) 5 and 6
9. (a) 16 (b) 6
   (c) 147 (d) 0.01
10. (a) 55 (b) 8100 (c) 200
11. No. $1^3 + 2^3 = 1 + 8 = 9$
    $3^3 = 27$
12. (a) 17 (b) 60
13. (a) $2^2 \times 3^2$ (b) $3^2 \times 5$
    (c) 9 (d) 180
14. 30 seconds
15. (a) 5 (b) 0.25
16. (a) $x = 9$ (b) $x = 3$
17. (a) 3^5 (b) 5^3 (c) 2^2
18. (a) 2 500 000 (b) 0.000 037
19. 3^4, $2^6 = 64$, $3^4 = 81$
20. 6.3
21. (a) 0.14
    (b) (i) 2.775 (ii) 2.7 (iii) 6.912
22. 4.25
23. (a) 47.1645… (b) 47.2
24. (a) 2.02
    (b) $\sqrt{3 + \frac{6}{3} - \frac{9}{3^2}} = \sqrt{3 + 2 - 1} = \sqrt{4} = 2$
25. (a) (i) 1.67 (ii) 26.62
    (b) $(1 + 2 + 3)^2 - (1^2 + 2^2 + 3^2)$
    $= 6^2 - (1 + 4 + 9)$
    $= 36 - 14$
    $= 22$

SECTION 7

Exercise **7**
Page 15

1. (a) 30% (b) 25% (c) 40%
2.

Fraction	$\frac{3}{4}$	$\frac{3}{10}$	$\frac{3}{5}$
Decimal	0.75	0.3	0.6
Percentage	75%	30%	60%

3. 0.02, 20%, $\frac{1}{2}$
4. (a) 2 pence (b) 15 kg (c) £45
5. Daisy. Daisy scored $\frac{4}{5} = 80\%$.

6. 16%
7. 1400
8. £15
9. £2.40
10. (a) 126
(b) 30%

11. (a) 20%
(b) 25%
12. £28
13. £630
14. 35%
15. £4.56

16. £29.75
17. £1.92
18. 25%
19. 36%
20. 81

21. Car A. Car A: $\frac{600}{3000} \times 100 = 20\%$

Car B: $\frac{2850}{15\,000} \times 100 = 19\%$

Exercise 8   Page 17

1. (a) 1453   (b) 19 minutes
2. (a) £16.43
(b) £10 + £5 + £1 + 2 × 20p + 2p + 1p
3. (a) 53 minutes   (b) 0739
4. (a) (i) 11.50 am, 1.10 pm
(ii) 1 hour 20 minutes
(b) £44.20
5. (a) 64 euros   (b) £6.25
6. (a) £171.50   (b) £45.83
7. £213.75
8. Small bar. Large: 2.66 g/p, small: 2.78 g/p
9. £480
10. £333.70

SECTION 9

Exercise 9   Page 19

1. £16.25
2. £1876.25
3. £596
4. £40.74
5. £8 × 40 = £320
6. (a) £1200
(b) £120
7. £267.38
8. £840

9. £6.20 per hour
10. £54 000
11. Missing entries are:
£22.50, £68.60,
£12.00, £80.60
12. £75 per month
13. £1725.90
14. (a) £69.48
(b) 1219 units

SECTION 10

Exercise 10   Page 21

1. (a) 1 : 3 (b) 2 : 1
(c) 2 : 3
2. 3 cm by 4 cm
3. 8 large bricks
4. 7.5 kg
5. (a) $\frac{1}{4}$ (b) 75%
6. 32 chocolates
7. 4 : 1
8. 840 males
9. £87

10. 12 women
11. 250 disabled people
12. 25 rock cakes
13. £1.96
14. £517.50
15. (a) 22.4% or 22%
(b) 1 : 7
16. 1 : 20 000
17. Year 9: 360
Year 10: 210
Year 11: 180

Exercise 11   Page 23

1. 64 km/h
2. $1\frac{1}{2}$ hours
3. 165 km
4. 40 minutes
5. (a) 90 mph
(b) $4\frac{1}{2}$ hours
6. 2 km
7. 1 hour 36 minutes
8. 16 miles per hour
9. (a) $2\frac{1}{2}$ hours
(b) 8 mph
10. 3 km/h

11. 0924
12. (a) 40 miles per hour
(b) 1116
13. Yes.
$\frac{65}{80} \times 60 = 48.75$ mins.
Arrives 1029.
14. 28.8 mph
15. 28.8 km/h
16. 10 m/s
17. 9 g/cm³
18. 19 g
19. 259.3 people/km²

Number

Non-calculator Paper   Page 25

1. (a) (i) 6, 10, 16, 61, 100 (ii) 193
(b) (i) 63   (ii) 2000   (iii) 25
2. (a) One million   (b) 3 tens, 30
3. (a) 2036
(b) (i) 0.5   (ii) 25%
(c) (i) 640   (ii) 600
4. (a) 3 and 5   (b) 3 and 9
5. (a) (i) 75   (ii) 133   (iii) 286
(b) 121   (c) (i) 2   (ii) 10
6. (a) (i) 6   (ii) 24   (iii) 5
(b) 1, 2, 3, 4, 6, 12
(c) 6
7. (a) 16 pence   (b) £1.88
8. 7 hours 47 minutes
9. E.g. 7 × 8 = 56
10. £16
11. (a) 74 pence   (b) £4.90
12. £4752
13. 8 minibuses
14. (a) 3569   (b) 9536
15. 0.83 kg
16. 1500 m
17. £30
18. 12 footballs
19. (a) £1.50   (b) £10.91
20. (a) 34.8   (b) 1.89
21. (a) 8°C   (b) 12
22. 60 pence × 90 = £54
23. 94 cm
24. (a) (i) 100 000   (ii) 68
(iii) 72   (iv) 0.9
(b) 5⁴, 5⁴ = 625, 4⁵ = 1024
(c) 50
25. (a) $\frac{7}{10}$ (b) £23.40   (c) (i) 0.08   (ii) 80

26. (a) $\frac{1}{2}$, $\frac{3}{5}$, $\frac{5}{8}$, $\frac{2}{3}$, $\frac{3}{4}$

    (b) $\frac{9}{40}$

    (c) (i) $\frac{13}{20}$ (ii) $\frac{1}{6}$ (iii) $\frac{8}{15}$

    (d) $4\frac{4}{5}$ or 4.8

27. (a) 8100   (b) 36 000

28. (a) 34.7   (b) $50 \times 300 = 15\,000$

29. 9

30. (a) £42.75   (b) £34.20

31. 99 pence

32. (a) £1.20   (b) 63p   (c) 90p

33. (a) 2 minutes 30 seconds

    (b) 2300 letters per hour

34. £3.60

35. (a) £11.40   (b) (i) 745   (ii) 754

36. (a) 16 km/h   (b) 1106

37. (a) 85%   (b) 20

38. (a) 29 and 37   (b) 200   (c) $\frac{7}{20}$

    (d) 47.3   (e) $\frac{30 \times 6}{0.3} = 600$

39. 75 mph

40. (a) 3^7 (b) 3^5 (c) 3^2

41. $4\frac{14}{15}$ metres

42. (a) 12.5 km   (b) 109.5 minutes

43. 10

44. (a) (i) $2^4 \times 3$ (ii) $2^2 \times 3^3$ (b) 432

Number

Section Review

Calculator Paper   Page 28

1. (a) 5   (b) 3570   (c) Four tenths

2. (a) -7, -1, 0, 5, 13   (b) 20

3. £2.35

4. £1.17

5. (a) 4.95 m

    (b) 4.95 m, 5.02 m, 5.10 m, 5.15 m, 5.20 m

6. (a) 0900   (b) 44 minutes

7. (a) £1861   (b) £186.10

8. Sue. Beth: 4, John: 12, Sue: 13

9. (a) 0.78   (b) 0.3, $\frac{8}{25}$, 33%, $\frac{1}{3}$

10. 2.8 pence

11. £281.60

12. 20

13. (a) 62.5   (b) 7.2   (c) 6

14. (a) 19 pence   (b) 68.2 g

15. (a) (i) £579   (ii) £10 229

    (b) (i) £9120   (ii) £760

16. 24.1 kg

17. England by £2.90

18. (a) 0.65   (b) 65%

19. 50

20. 60 km/h

21. (a) 27.24040786   (b) 30

22. 0.5 litres

23. (a) 474 units   (b) £55.95

24. E.g. $2^2 + 3^2 = 4 + 9 = 13$, **not** even.

25. Small.

    Small: $\frac{180}{36} = 5$ g/p   Large: $\frac{300}{63} = 4.76$ g/p

26. £77

27. 20 cards and 32 cards

28. £104.59 (or £105)

29. £115.15

30. 0909

31. 12.5%

32. (a) (i) $2^3 \times 3^2$ (ii) $2^4 \times 5$

    (b) 12 minutes

33. 889.61 dollars

34. (a) 4   (b) 51.2

35. (a) £227.50   (b) £167.44

36. (a) $\sqrt{6.9}$, 2.58, $2\frac{4}{7}$, 1.6^2

    (b) (i) 290 (ii) $\frac{600 \times 30}{80 - 20} = \frac{18\,000}{60} = 300$

37. 3150

SECTION 12

Exercise 12   Page 31

1. £9k

2. $(t + 5)$ years

3. (a) $4x$ (b) $3x + 7y$ (c) $12a$

4. $(3x + 2y)$ pence

5. (a) $6m$ (b) $m + 2$ (c) m^3

6. (a) $4x$ pence   (b) $4m - 1$

7. $(4x + 200)$ degrees

8. (a) $n + 1$ (b) $4n + 2$

9. $(5d + 15)$ pence

10. (a) $8p$ (b) t^3 (c) $3x + 4y$

11.

$2y$	and	$y + y$
y^2	and	$y \times y$
$2(y + 1)$	and	$2y + 2$
$2y + y$	and	$3y$

12. (a) $x + 2y$

    (b) (i) $2x + 6$ (ii) $x^2 - x$

13. (a) $10a^7$ (b) $6gh$

    (c) $2k$ (d) 3

14. (a) $(x + 3)$ years

    (b) $(2x + 6)$ years

15. (a) $7p + q$

    (b) (i) $4r - 12$ (ii) $s^3 + 6s$

16. (a) £xy (b) £$y(x - 5)$

17. (a) $7a + 2b + 4ab$ (b) $4x + 15$

18. $14x - 9$

19. (a) $15p - 3q$ (b) $18x + 44$

20. (a) y^5 (b) x^3 (c) z^2

21. (a) $3(x - 2)$ (b) $x(x - 2)$

22. (a) (i) $8 - 6n$ (ii) $4x - 1$

    (b) $p^2 + p - 12$

23. (a) a^4 (b) b^4 (c) c^3 (d) d^4

24. (a) $2(2x + 3)$

    (b) (i) $6y - 9$ (ii) $x^3 - 2x^2$ (iii) $a^2 + ab$

    (c) $x^2 - x$ (d) $m^2 - 5m + 6$

1. (a) 15   (b) 9   (c) 5   (d) 15
2. (a) $x = 5$ (b) $x = 2$
   (c) $x = 14$ (d) $x = 7$
3. (a) 55   (b) 7
4. (a) 5   (b) 4
5. (a) $x = 10$ (b) $x = 5$
   (c) $x = 4$ (d) $x = -6$
6. (a) $x = -1$ (b) $x = \frac{1}{2}$
   (c) $x = 5\frac{1}{2}$ (d) $x = -0.8$

1. (a) $x = 4$ (b) $x = -1$
   (c) $x = 5$ (d) $x = 21$
2. 11
3. $w = 6$, $x = 8$, $y = 5$, $z = 4$
4. (a) $x = 11$ (b) $x = 6$
5. (a) $x = 8$ (b) $x = 1$
   (c) $x = -4$ (d) $x = 2.5$
6. (a) $x = 6$ (b) $x = 3$ (c) $x = 1.5$
7. (a) $x = -1.5$ (b) $x = 2.5$
   (c) $x = 0.6$ (d) $x = 1.5$
8. $y = \frac{1}{2}$
9. (a) $n + (n + 3) + (2n - 1) = 4n + 2$
   (b) $4n + 2 = 30$, $n = 7$
10. $n + (2n + 5) = 47$, $3n + 5 = 47$, $n = 14$.
    Larger box has 33 chocolates.
11. (a) $(40y + 180)$ pence   (b) 7 cakes
12. $x = -2$
13. (a) $x = 7$ (b) $x = 0.6$
14. $x = 5$
15. $x = 1.5$, $A = 7$
16. (a) $x = -21$ (b) $x = 2$
    (c) $y = -3.5$ (d) $z = -0.8$
17. (a) $x = 17$ (b) $x = 1.2$
18. (a) $x = 5$ (b) $y = 12$ (c) $z = 5$
19. $x = 4.8$

1. 4
2. -1
3. (a) 39   (b) (i) 28   (ii) 4
4. (a) 2   (b) -8 (c) 8   (d) -15
5. $H = -13$
6. (a) -1 (b) 3
7. (a) £17   (b) 4 hours
8. $L = -10$

9. $A = -11$
10. (a) -60 (b) 6
11. (a) 6   (b) 16
12. 90
13. 24
14. $T = 100$
15. (a) £2.35   (b) 12 miles
16. (a) 111.8   (b) 140
17. (a) 40 km   (b) $K = \frac{8M}{5}$
    (c) $M = 37.5$
18. 59°F
19. (a) $-1\frac{5}{8}$ (b) $-\frac{5}{8}$
20. $t = \frac{c + 5}{3}$
21. (a) $c = \frac{P - 2a - 2b}{2}$ (b) $c = 4.5$
22. $d = 62.8$
23. $n = 14$
24. $r = \frac{ps}{g}$
25. (a) $v = -7$ (b) $a = \frac{v - u}{t}$

1. (a) 21, 25   (b) 30, 26
2. (a) 17   (b) 81   (c) $\frac{1}{16}$
3. 37, 60
4. (a)
   Pattern 4
   (b) Missing entries are: 7, 9
   (c) 11
   (d) 35
5. (a) 14
   (b) No. Number must be (multiple of 3) $- 1$.
6. 5, 6, $5\frac{1}{2}$
7. (a) Multiply the last term by 3.   (b) 405
8. -2, -8
9. (a) -10
   (b) Subtract the next even number (-8).
10. No. The sequence does not end.
    Sequence: 1, 6, 10, 8, 4, 8, 8, 0, 16, …
11. (a) Pattern 20 has 58 squares.
    $3 \times$ (pattern number) -2
    (b) $3n - 2$
12. (a) 40. Add 7 to the last term.   (b) $7n - 2$
13. (a) $2n + 3$ (b) $4n - 3$
14. (a) 5, 8, 13
    (b) No.
    $102 (= 106 - 4)$ is not a square number.

1. (a) $R(-6, 2)$, $S(3, -4)$
   (b) $T(-3, 0)$, $U(0, -2)$

2. (a)

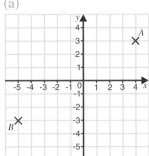

(b) $p = -2$

3. (a)

(b) $(5, 2)$

4. (a) $P(0, 3)$
$Q(6, 0)$
(b) $m = 5.5$

5. (a) $y = 3$
(b) $y = \frac{1}{2}x + 1$

6. (a) Missing entries are: $7, -5$

(b)

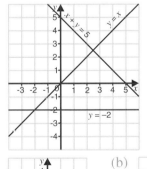

(c) $y = 4$

7.

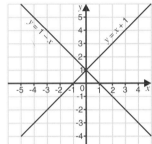

8. (a)

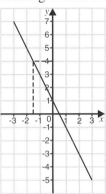

(b) $x + y = 5$

9. (a) (b)

(c) (d)

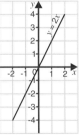

10. (a) Missing entries are: $-6, 3$.

(b)

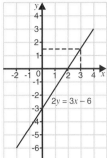

(c) $x = 3$

11. (a)

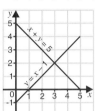

(b) $y = 1.2$

12. (a)

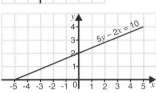

(b) $(3, 2)$

Exercise **18**

1. (a) 5.6 km (b) 4.4 miles (c) 20.8 km
2. (a) 8°C (b) 30 minutes (c) 30 minutes
3. (a)

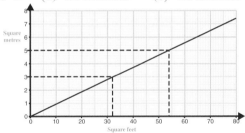

(b) (i) 54 square feet (ii) 3 square metres
4. (a)

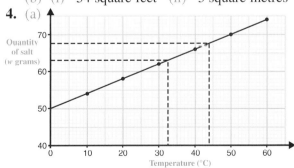

(b) (i) 32.5°C (ii) 67.6 grams
(c) (i) $a = 0.4$, $b = 50$ (ii) 88 grams
5. (a) £480 (b) £100 (c) £1120
6. (a) Decreased (b) 70 minutes (c) 8 mph
7. (a) 30 minutes (b) 18 km (c) 36 km/h
8. (a) 1016 (b) 28 km
(c) Between B and C. Steepest gradient.
(d) 50 km/h

9.

Exercise 19 — Page 45

1. (a) $x > 3$ (b) $x \geq -2$
    (c) $x \leq 6$ (d) $x > 2$

2. (a)

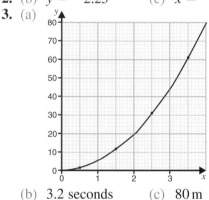

(b)

(c)

(d)

3. (a) $x \leq 3$
    (b)

4. (a) $x \leq 2$ (b) $x > 3\frac{1}{2}$ (c) $x < -\frac{4}{3}$

5. (a) $-1, 0, 1, 2$ (b) $1, 2$ (c) $-1, 0, 1$

6. $2, 3$

7. (a) $x \geq -1$ (b) $x < 2$ (c) $-1, 0, 1$

Exercise 20 — Page 46

1. (a) Missing entries are: $7, -1, 2$
    (c) $x = \pm 2.2$ (d) $x = \pm 1.4$

2. (b) $y = -2.25$ (c) $x = -2$ or 1

3. (a)

(b) 3.2 seconds   (c) $80\,\text{m}$

Algebra

Non-calculator Paper — Page 47

1. $A\,(4, 1)$

2. (a) 8 (b) Not an even number.

3. (a) 48 (b) 7

4. 26 points

5. (a)

    (b) Missing entry is: 14
    (c) 17 sides
    (d) 35 sides

6. 21

7. (a)

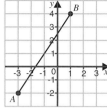

    (b) $(-1, 1)$

8. (a) $5t$ pence   (b) $(t + 5)$ pence

9. (a) 10 (b) 4 (c) 2 (d) 15

10. (a) (i) 30
      (ii) 50th term is an odd number.
        All even terms are odd numbers.
    (b) Add 4 to the last term.

11. (a) **could be even or odd**   (b) **always odd**

12. 10

13. $(3x + 5y)$ pence

14. 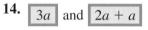 $3a$ and $2a + a$ $2(a - 1)$ and $2a - 2$

15. (a) $5x$ (b) $3a - 4b$ (c) $3m^2$

16. (a) $\frac{1}{2}$ (b) -1

17. (a) $2n + 5$ (b) 2

18. (a) 6 (b) 12

19. (a) $x = 4$ (b) $g = 3$
    (c) $y = \frac{1}{2}$ (d) $y = -1$

20. $P\,(-1, -3)$

21. $80\,\text{mph}$

22. -11

23. (a) $(x - 3)$ years   (b) $4x$ years
    (c) $x + (x - 3) + 4x = 45$, $x = 8$.
      Louisa 5 years,   Hannah 8 years,
      Mother 32 years

24. (a) (i) $3(a - 2)$ (ii) $k(k - 2)$
    (b) (i) $5x + 15$ (ii) $m^2 - 4m$
    (c) (i) $x = -1$ (ii) $x = \frac{1}{2}$

25. $t = 5$

26. $x = 1.5$

27. (a) $x = 3$ (b) $x < 5$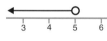

28. (a)

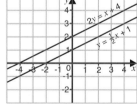

    (b) Lines are parallel, same gradient.

29. (a) $(5x + 15)$ pence
    (b) $x + 5x + 15 = 87$, $x = 12$.
      Pencil costs 12p.

30. $4n + 1$

31. (a) $x < 1\frac{1}{2}$ (b) $-1, \ 0, \ 1$

32. **A: Q,   B: S,   C: R,   D: P**

33. (a) $x = -3.5$ (b) y^8

34. (a) $3n - 5$ (b) $x = \dfrac{y + 5}{2}$

35. (a) Missing entries are: $4, 1$
    (c) (i) $x = 1$ (ii) $x = -0.4$ or 2.4

36.

Depth / Time graph

37. (a) $p^2 - 4$
(b) q^6

Algebra

Calculator Paper — Page 50

1. (a) (i) 22
(ii) Add the next counting number, 6.
(b) 12
2. (a) (i) $6y$ (ii) $4m + 1$ (b) (i) 14 (ii) 8
3. (a) $3x$ pence (b) $(x + 30)$ pence
4. (a) £325 (b) 11 cars
5. (a) (i) 11, 13 (ii) 32, 64 (b) 43, 77
6. (a) Missing entries are: $3a$, $9a + 2b + 6c$
(b) 57
7. (a) (i) 25 dollars (ii) 18 euros
(b) From graph, 20 dollars = 24 euros.
So, 100 dollars = $5 \times 24 = 120$ euros.
8. No. $6 + 16 = 22$. Then, $22 \times 2 = 44$.
9. (a) Missing entries are: -3, 1
(b)

(c) $(0, -2)$, $(2, 0)$

10. (a) $g = 8$ (b) $a = 5$ (c) $x = 6$ (d) $x = 3$
11. (a) $C = 4u$ (b) $u = 7$
12. (a) 7 (b) $m = 3$ (c) $P = 45$
13. (a) 6 (b) 17
14. (a) $2x + 1 = 35$ (b) 17
15. (a) 30 minutes (b) 16 mph
(c) From C to D, steepest gradient.
(d) 4 pm or 1600
16. $F = 3S - 4$
17. (b) $P(-5, -9)$
18. (a) (i) $x = 20$ (ii) $x = \frac{1}{2}$ (b) $3t - 12$
19. (a) 2, 5 (b) 11th term
(c) 85 is not (a multiple of 3) $- 1$
20. (a) (i) $a = 3.5$ (ii) $t = 1$
(b) $x + x - 3 + x + 7 = 25$
$3x + 4 = 25$, $x = 7$.
21. (a) 7 (b) (i) $x + 3$ (ii) $3n - 1$
22. $x = 2.6$, correct to 1 d.p.
23. (a) $(x + 45)$ pence
(b) $3(x + 45) + x = 455$, $x = 80$.
Glass of milk costs 80 pence.
24. (a) Missing entries are: -1, -4, -1
(c) $x = \pm 2.2$
25. $x = 4.3$, correct to 1 d.p.
26. (a) $14x - 13$ (b) $x^2 + 6x + 8$
27. (a) $x = -\frac{1}{5}$ (b) $7x - 3$ (c) (i) m^6 (ii) n^5
28. $x = \dfrac{7 - y}{3}$
29. (a) 1, 2, 3 (b) $y(x - y)$

SECTION **21**

Exercise 21 — Page 54

1. (a) CD and EF (b) AB and CD
(c) (i) $y = 135°$ (ii) **obtuse angle**
2. (a) $70°, 80°$ (b) $100°, 130°$ (c) $250°$
3. (a) $a = 143°$ (supplementary angles)
(b) $b = 135°$ (angles at a point)
(c) $c = 48°$ (vertically opposite angles)
$d = 132°$ (supplementary angles)
$e = 44°$ ($3e = 132°$, vert. opp. angles)
4. (a) $x = 53°$ (alternate angles)
(b) $y = 127°$ (allied angles or
supplementary angles)
5. (a) $x = 48°$ (corresponding angle, $\angle QPR$)
(b) $y = 97°$
6. $a = 68°$ (supplementary angles)
$b = 112°$ (corresponding angles)
$c = 106°$ (allied angles)
7. (a) $a = 105°$ (b) $b = 117°$, $c = 117°$
(c) $d = 42°$, $e = 76°$, $f = 62°$

SECTION **22**

Exercise 22 — Page 55

1. (a) $a = 27°$ (b) $b = 97°$ (c) $c = 125°$
2. $\angle ACB = 73°$
3. $x = 68°$ ΔPQR is isosceles, $\angle PQR = \angle PRQ$
4. (a) $x = 36°$ (b) $y = 108°$
5. $\angle CDA = 30°$
$\angle BCA = 60°$ (ΔABC equilateral)
$\angle ACD = 180° - 60° = 120°$ (supp. $\angle$'s)
$\angle CDA = \frac{1}{2}(180° - 120°)$
$= 30°$ (ΔACD isosceles)
7. (a) $9\,cm^2$ (b) $10\,cm^2$ (c) $13.5\,cm^2$
8. (b) $15.2\,cm^2$
9. $27\,cm^2$
10. $YX = 4\,cm$

SECTION **23**

Exercise 23 — Page 58

1.

4.

5. (a) 4 (b) 1
6. **A** and **F**
2. (a) A, E (b) N
(c) O
7. (a)
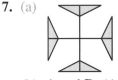

3. (a) (i) 3 (ii) 3
(b) (i) 0 (ii) 1
(c) (i) 0 (ii) 4
(d) (i) 1 (ii) 1
(b) **A** and **D** (ASA)

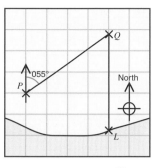

6. (a) (i) 124°
(ii) 304°
(b) 6.25 km

7. (b) (i) 250°
(ii) 1530 m

SECTION 24

Exercise 24 — Page 60

1. (a) 14 cm²
(b) (i)

Not full size

(ii) 8 cm², 18 cm², 20 cm²
2. (a) Rhombus (b) Parallelogram
(c) Trapezium
3. (a) $a = 70°$ (b) $b = 132°$
(c) $c = 110°$, $d = 120°$
4. (a) **kite** (b) $\angle ABC = 114°$
5. 6 cm
6. (a) 30 cm (b) 55.04 cm²
7. $b = 6$ cm
8. (a) $x = 51°$ (b) 40 cm²

SECTION 25

Exercise 25 — Page 61

1.

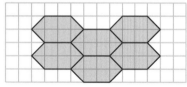

2. Shape A has rotational symmetry of order 6
and 6 lines of symmetry.
Shape B has rotational symmetry of order 2
and 2 lines of symmetry.
3. (a) $a = 53°$ (b) $b = 115°$
(c) $c = 140°$
4. (a) $a = 120°$ (b) $b = 60°$, $c = 120°$
(c) $d = 72°$, $e = 108°$
5. 15 sides
6. A hexagon can be divided into
4 triangles, as shown.

Sum of the interior angles in a triangle = 180°.
Sum of interior angles of a hexagon is
$4 \times 180° = 720°$.
7. (a) $x = 40°$ (b) $y = 140°$
8. (a) $\angle ABC = 144°$ (b) $\angle XCY = 108°$
9. $\angle LMX = 141°$

SECTION 26

Exercise 26 — Page 63

1. South-East
2. (a) Glasgow (b) 225°
3. 26 cm
4. (a) 055° (b) 345° (c) 235°

SECTION 27

Exercise 27 — Page 66

1. 6 cm
2. (a) 18.8 m
(b) 28.3 m²
3. (a) 22 cm
(b) 38.5 cm²
4. 225π cm²
5. 754 cm²
6. 19.9 times
7. (a) 346 cm
(b) 7977 cm²
8. 58.1 cm²
9. 106 revolutions
10. (a) 220 cm
(b) 47.7 cm
11. 16π cm²
12. 796 cm²
13. Yes. Semi-circle $= \frac{1}{2}(\pi \times 10^2) = 50\pi$ cm²
Circle $= \pi \times 5^2 = 25\pi$ cm²

SECTION 28

Exercise 28 — Page 68

1. (a) 5 faces, 8 edges, 5 vertices (b) **R**
2. (a) (b)

3. (a) 20 cm
(b) 9 cm²
(c) Any shape with area 9 cm².
4. 13 cm²
5. (a) 400 cm (b) Length 60 cm, width 20 cm
6. (a) 30 cubes (b) (i) (ii) 62 cm²

7. (a) **C**.
A = 24 cm³, **B** = 24 cm³, **C** = 27 cm³
(b) (i) (ii) 52 cm²

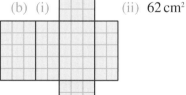

(c) **B**: 56 cm²
8. (a) (i) 72.25 cm² (ii) 78.54 cm²
(b) Yummy. Volume = base area × height
Yummy has larger base area.

9. 8.6%, correct to 1 d.p.

10. 14 000 cm³ (0.014 m³)

11. 297 m²

12. (a) 414 cm²

(b) 405 cm³

13.

14. 151.3 cm³

15. 2.5 cm

16. No. $h = \dfrac{1000}{\pi \times 7^2} = 6.5$ cm

Total depth = 9 + 6.5 = 15.5 cm

15.5 cm < 18 cm

17. (a) 32 673 cm²   (b) 20.1 cm

SECTION 29

Exercise 29 — Page 72

1.

2.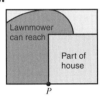

3. (a)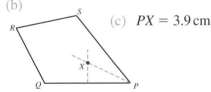

(b) 8.4 km

4. (a) (b)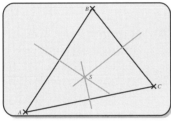

(c) $PX = 3.9$ cm

SECTION 30

Exercise 30 — Page 74

1. (a) 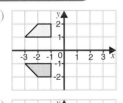 (b)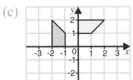

(c)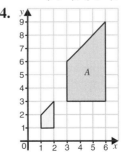

2. (a) Rotation, 90° anticlockwise, about (0, 0).

(b) Translation 4 units to the right and

3 units down $\left(\text{or } \begin{pmatrix} 4 \\ -3 \end{pmatrix}\right)$.

(c) Reflection in $y = 3$.

3.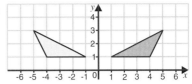

4. Rotation, through 180°, about (0, 0).

5. (a) $x = -1$

(b) One unit to the left (c)
and 2 units up.

6. (a) Reflection in $y = 2$.

(b) (i) 180°   (ii) (0, 3)

(c) Coordinates of triangle C:
$(2, -1), (5, -1), (5, -3)$.

7. (a) Reflection in $x = 3$.

(b) Rotation, through 180°, about (2, 1).

(c) Translation $\begin{pmatrix} 2 \\ -3 \end{pmatrix}$.

8. (a) Rotation, 90° anticlockwise, about (0, 0).

(b) Reflection in $y = x$.

9. $(-4, -1)$

10. (a) $Q(1, 3)$

(b) (i) $S(4, -2)$ (ii) $\begin{pmatrix} -1 \\ 3 \end{pmatrix}$

(c) $X(1, -1)$

SECTION 31

Exercise 31 — Page 77

1.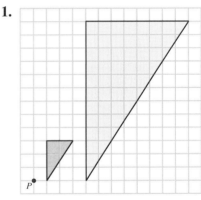

2. Enlargement, scale factor 3, centre (0, 0).

3. (a) Centre (0, 2), scale factor 2.

(b) Coordinates of enlarged shape:
$(3, 2), (3, 4), (1, 4), (-1, 0)$

4.

5. (a) Ratio of widths = 1 : 2 **but**
ratio of lengths = 5 : 7.

(b) 7.5 cm

6. (a) $XY = 10$ cm   (b) $AC = 1.2$ cm

SECTION 32

Exercise 32 — Page 78

1. $BC = 13\,cm$
2. 6 km
3. 28.7 cm
4. 9.85 units
5. $QR = 7.48\,cm$
   area $\Delta PQR = 18.7\,cm^2$
6. (a) 19.2 km
   (b) 9.8 km
7. (a) 13 cm²
   (b) $AB = 5.4\,cm$
8. 63.4 cm²

SECTION 33

Exercise 33 — Page 81

1. (a) metres
   (b) grams
   (c) litres
2. (a) 1.4 m
   (b) 8500 g
3. $\frac{1}{4}$
4. (a) 28 kg
   (b) 85 km/h
   (c) 2.65 grams
5. (a) 6 km
   (b) 60 g
   (c) 60 cm²
   (d) 60 m³
6. (a) 40 000 m
   (b) 25 miles
7. (a) 1.35 m²
   (b) 200 000 cm³
8. Door A.
   $2\,m = 200\,cm$
   70 inches $= 70 \times 2.5$
   $= 175\,cm$
9. £813
10. Greatest: 20.5 kg
    Least: 19.5 kg
11. 16 km/h
12. 2.45 kg
13. $\frac{abc}{2}$

Shape, Space, Measures — Section Review

Non-calculator Paper — Page 82

1. (a) $AB = 7.2\,cm$ (b) (c)

2. (a) 12
   (b) 5
   (c) cone
3. (a) 22 cm²
   (b) (i) 12 cm³
       (ii)   (iii)

4. (a) $x = 34°$, $y = 110°$ (b) y
5. (a) $a = 58°$ (vertically opposite angles)
   (b) $b = 155°$ (supplementary angles)
   (c) $c = 150°$ (angles at a point)
6. (a) (i) Equilateral (ii) Hexagon
   (b) (i) 3   (ii) 3
7. (a) Reflection in the x axis.
   (b) (i) 2   (ii) BC and FG, CD and GH,
                   DE and HI.
8. (a) 4.8 kg
   (b) More. $10\,lb = \frac{10}{2.2}\,kg = 4.5\ldots\,kg$

(Section 33 continued — right column)

9. (a) $a = 67°$ Angles in a triangle add to 180°.
   (b) $b = 54°$ Isosceles Δ.
       $b = 180° - (2 \times 63°)$
   (c) $c = 126°$ Angles in a quadrilateral add
                    to 360°.
       $c = 180° - (360° - 306°)$
10. (a) **trapezium**   (b) **rhombus**
11. (a) 18 cm²   (b) 25 cm²
12. $30 \times 20 \times 10 = 6000\,cm^3$
13. **B** and **E**   (ASA)
14. (a) 060°   (b) 300°
15. (a) $25\pi\,cm^2$ (b) $(10 + 10\pi)\,cm$
    (c) (i) $w = \sqrt{50}\,cm$ (ii) $h = 10\,cm$
16.
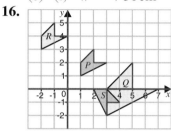
17. 140°   18.
19. 3.6 cm
20. $14\pi\,cm$
21. Greatest: 30.5 m
    Least: 29.5 m
22. (a) abc has dimension 3 (volume)
    (b) πa and $\sqrt{a^2 - c^2}$ and $2(a + b + c)$
23. 45 cm³

Shape, Space, Measures — Section Review

Calculator Paper — Page 85

1. (a) 8000 g and 8 kg   (b) 0.2 km   (c) 1.4 kg
2. (a) $\frac{3}{8}$ (b) North
3. (d) $\angle APB = 90°$
4.
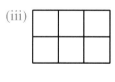
   mirror line
5. (a) 16 cm   (b) G
   (c)
6. (a) $A\,(2, 3)$ (b) $C\,(-2, -1)$ (c) $D\,(-2, 3)$
7.
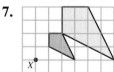
8. (b) $\angle PQR = 120°$
   (c) Obtuse
9. Height 175 cm
   Weight 70 kg
10. (a) Reflection in the line $x = 5$.
    (b) Rotation, 90° anticlockwise, about (0, 0).
    (c) Translation $\begin{pmatrix} 4 \\ 1 \end{pmatrix}$
11. (a) $a = 40°$, $b = 95°$, $c = 135°$ (b) $d = 40°$
12. 3.6 cm
13. (a) E.g.   (b) E.g.

14. (a) E.g.

(b) No. Only equilateral triangles, squares and hexagons tessellate. Interior angle must divide into $360°$ a whole number of times.

15. (a) (i) 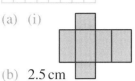   (ii) $46\,cm^2$

(b) $2.5\,cm$

16. $28.3\,cm$

17. E.g. $a = 4\,m$, $b = 6\,m$; $a = 3\,m$, $b = 7\,m$; etc. $(a + b = 10)$

18. (a) Ext. $\angle = \frac{360°}{6} = 60°$.

$a = $ int. $\angle = 180° - 60° = 120°$

(b) $b = 15°$

19. $36.25\,km$

20. $(2.5, -1)$

21. (a) $540\,cm^2$ (b) $AC = 39\,cm$

22. $769\,690\,cm^3$

23. (b) (i) $215°$ (ii) $17\,km$

24. Footpaths HX, XS shorter by $171\,m$. Footpaths $850\,m$, Waverly Crescent $1021\,m$.

25.

SECTION 34

Exercise 34   Page 88

1. (a) (i) E.g.

Gender	Mode of transport
M	Bus
F	Cycle

(b) Data will be biased, as people at bus station are likely to travel by bus.

2. (a) Laila   (b) Ria

(c) Corrin. Pupils arc in thc samc class and June is a later month in the school year than thc other months givcn.

3. E.g.

Resort	Tally
Cervinia	HHT II
Livigno	HHT II
Tonale	HHT IIII

4. (a) E.g. May play a different sport. May play more than one of these sports. Answer does not indicate which sport is played.

(b) Which sport(s) do you play?

Football ☐   Rugby ☐   Hockey ☐

Netball ☐   Swimming ☐   Tennis ☐

Other ☐   (state) …………   None ☐

5. (a)

Time (t minutes)	Tally	Frequency
$0 \leqslant t < 30$	HHT	5
$30 \leqslant t < 60$	HHT HHT II	12
$60 \leqslant t < 90$	HHT II	7
$90 \leqslant t < 120$	HHT	5
$120 \leqslant t < 150$	I	1

(b) $30 \leqslant t < 60$

(c) E.g. More homework may be set on a Wednesday night. Homework may take more (less) time than usual.

6.

	Red	Yellow	Green
Male	3	2	2
Female	2	1	3

7. (a) Leading question.

(b) Biased, as students from Wentbridge are likely to agree with the question.

8. (a) (i) Too personal.

(ii) In which age group are you?

Under 16 ☐   16 to 19 ☐   Over 19 ☐

(b) (i) Only students already using the library are sampled.

(ii) Give to students as they enter (or leave) the college.

9. No.

Men: $\frac{180}{200} = 90\%$

Women: $\frac{240}{300} = 80\%$

Higher proportion of men can drive.

10. E.g. Two thirds of men were over 45. All women are aged 16 to 45. Twice as many women as men.

11. E.g. Only a small sample. People in favour are less likely to write in. People could be objecting to the amount of money offered, **not** the proposal.

12. (a) 6   (b) 10

(c) 7   (d) 28

SECTION 35

Exercise 35   Page 91

1. (a)

(b) 20%

2. (a) Jack   (b) 15

3. (a) 12   (b) 34

(c)

Football	🏃🏃🏃	
Rugby	🏃	🏃 = 4 friends
Racing	🏃🏃	
Other	🏃🏃🏃	

4. (a) Wednesday   (b) 3 hours   (c) 39 hours
5. (a) Range: 2,   mode: 2
    (b) Women's team had a larger range (5) and higher mode (3).

SECTION 36

Exercise 36   Page 94

1. (a) 7 cm   (b) 10 cm
    (c) 11 cm   (d) 11.3 cm
2. (a) £9   (b) £10.50   (c) £11.50
    (d) Median.
       Mode is the lowest price and mean is affected by the one higher-priced meal.
3. (a) 14.275
    (b) All values have increased by 1, so, add 1 to mean.
    (c) 14.6
    (d) Mean.   The highest value has been increased, so, median would not change, but mean would increase.
4. (a) £200   (b) £350   (c) £420
    (d) E.g. The advert refers to the average wage of **all** employees. A mechanic only earns £250 per week.
5. (a) (i) 4.9 hours   (ii) 6 hours
    (b) Much bigger variation in the number of hours of sunshine each day and lower average.
6. (a) 3   (b) 1.92   (c) Mode
7. (a) **B**   (b) 12   (c) 1.9
8. (a) 30.5 minutes   (b) $25 \leqslant t < 30$

SECTION 37

Exercise 37   Page 96

1. (a) 12   (b) 22°C   (c) 14°C
    (d) 5°C or 24°C.   Temperature can be either 2°C above previous maximum, or 2°C below previous minimum.
2. (a) 18   (b) 72
3. (a)

Fuel	Solid fuel	Electricity	Gas
Angle	24°	126°	210°

    (b) (i) $\frac{5}{36}$ (ii) 72
    (c) Gas.   Town has larger population and proportion of gas users is greater than 50%

4. (a)

1	0 means 10 text messages
0	2 3 5 5 7 9
1	0 1 2 3 5 7
2	0 1

    (b) 19
5. (a)

Destination	France	Spain	Italy	Greece	America
Angle	72°	108°	36°	84°	60°

    (b) 20%
6. (a)

Boys		Girls 2	5 means 2.5 cm
	2	5	
5 5	3	0 5 5 5	
5 5 5 5 0 0	4	0 5 5	
0 0	5	0 5	

    (b) Girls have more variation in their estimates than boys:
       Girls 3.0 cm, Boys 1.5 cm.

SECTION 38

Exercise 38   Page 99

1. (a)

    (b) 8°C
    (c) (i) 16°C
       (ii) Actual temperatures are only known at times when readings are taken.
2. (a) 20 minutes or more but less than 30 minutes.
    (b)

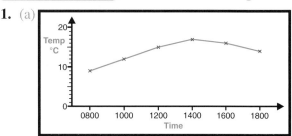

3.

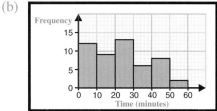

4. (b) 5%
5. Vertical scale is not uniform.

6. (a)

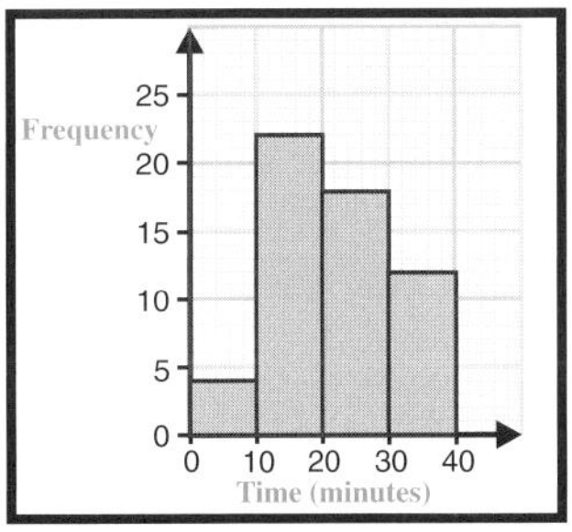

(b) $10 < t \leqslant 20$
(c) E.g. Responses overlap (e.g. £7 spent could be represented by a tick in first 3 boxes)

7. (a) $80 \leqslant age < 90$ (b) 40
(c) (ii) Women:

Age (a years)	Frequency
$60 \leqslant a < 70$	1
$70 \leqslant a < 80$	5
$80 \leqslant a < 90$	13
$90 \leqslant a < 100$	6

(iii) More men under 80 than women. Only women aged over 90. Women have greater range of ages.

8. (a) 7 (b) 3 (c) 67

SECTION 39

Exercise 39 Page 101

1. (b) Negative correlation. As engine size increases fuel economy decreases.
(d) Points are close to the line of best fit.
2. (a) (i) ***C*** (ii) ***A*** (iii) ***B*** (b) ***A***
3. (a) 36 (b) 60 (c) No
(d) Yes (e) Positive
4. (c) (i) 3.8 m (ii) 6.2 m
(d) (c)(i) as estimated value lies between known values.

SECTION 40

Exercise 40 Page 104

1. (a) (i) **evens** (ii) **likely** (iii) **impossible**
(b) The sum of the probabilities should be 1. $0.3 + 0.6 + 0.2 > 1$
2. (a)

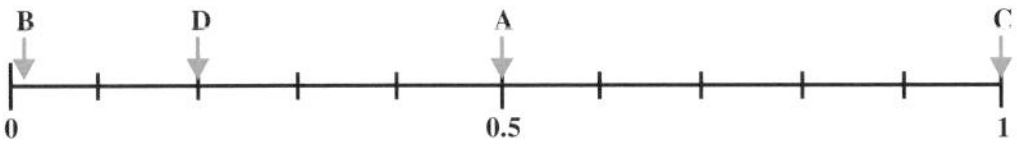

(b) 0.2 (c) 0.8
3. (a) $\frac{1}{8}$ (b) $\frac{4}{8} = \frac{1}{2}$ (c) $\frac{5}{8}$
4. (a) **1, 2** **1, 3** **1, 4**
2, 2 **2, 3** **2, 4**
3, 2 **3, 3** **3, 4**
(b) $\frac{2}{9}$
5. (a) $\frac{2}{5}$ (b) 0.6

6. (a)

		1st set of cards			
		1	**2**	**3**	**4**
2nd set of cards	**5**	6	7	8	9
	6	7	8	9	10
	7	8	9	10	11
	8	9	10	11	12

(b) (i) $\frac{1}{16}$ (ii) $\frac{6}{16} = \frac{3}{8}$ (c) 25
7. (a) $\frac{9}{20} = 0.45$
(b) 2, 3, 3, 4, 5. Numbers 2, 3, 4, 5 have occurred and 3 has occurred twice as often as other numbers.
(c) 100. Relative frequency of 5 is $\frac{1}{5}$. $\frac{1}{5} \times 500 = 100$
8. (a) (i) $\frac{11}{30}$ (ii) $\frac{17}{30}$ (b) $\frac{7}{11}$
9. (a) Missing entry is: 0.25
(b) Even. P(odd) = 0.45 P(even) = 0.55
10. 140

Handling Data — Section Review

Non-calculator Paper Page 106

1. (a) 4 (b) 11 (c) 22
2. (a)

Number of pets	Tally	Frequency
0	||||	4
1	~~||||~~ |||	8
2	~~||||~~	5
3	|||	3
4	||||	4
5	|	1

(b)

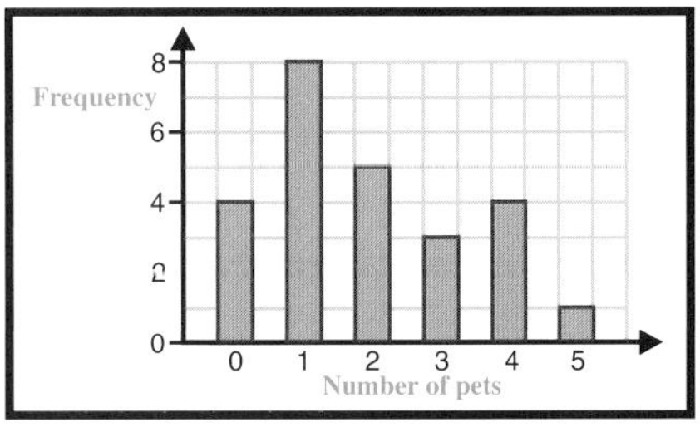

3. (a) Spain (b) 12 (c) 51
4. (a) 8 (b) 7.5 (c) 6 (d) 7
5. (a) 3 (b) 30 (c) 60
6. (a) (i)

Pantomime	Angle
Aladdin	135°
Cinderella	105°
Jack and the Bean Stalk	75°
Peter Pan	45°

(ii) Aladdin
(b) (i) 72 (ii) $33\frac{1}{3}\%$

7. (a)

	Second spin		
	Red	Yellow	Blue
First spin — Red	2	1	1
First spin — Yellow	1	2	0
First spin — Blue	1	0	2

   (b) $\frac{2}{9}$ (c) 23

8. E.g. Leading question.
Question has more than one part.

9. (a) Unequal intervals. Class intervals 3 to 6 and 6 to 8 overlap. No class greater than 8.
   (b) Anna. Larger sample size.

10. (a)

		4	5 means 4.5 cm
4	5 8 8		
5	0 0 4 4 5 8		
6	0 2 4 5 5 5 6 8		
7	0 2 4		

   (b) 2.9 cm

11. (a) $\frac{17}{75}$ (b) Yes. Female: $\frac{12}{50} = 24\%$
                  Male: $\frac{5}{25} = 20\%$

12. (c) (i) 120 hours   (ii) 35 hours
   (d) (c)(i) as it lies between known values.

13. (a) 0.7   (b) 0.44

Handling Data

Calculator Paper   Page 109

1. (a) 170
   (b)

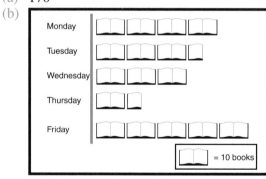

Monday	= 10 books
Tuesday	
Wednesday	
Thursday	
Friday	

2.

David   Barbara   Abdul   Cathy   Ewan

3. (a) 12 m   (b) 11 m   (c) 10.4 m
4. (a) 2   (b) 31   (c) 96
5. (a) **X 1**, **X 3**, **Y 1**, **Y 3**
   (b) Numbers 1 and 3 are not equally likely.
6. (a) 20   (b) 6   (c) 5
   (d)

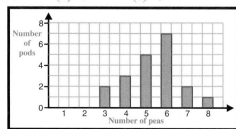

Number of pods (y-axis), Number of peas (x-axis)

---

7. (a) 15   (b) 28 g   (c) 30 g   (d) 29.3 g
8. (a) 51%
   (b) Missing entries are: 43%, 9%, 2%
   (c) (i) In school (ii) 25% (iii) $\frac{60°}{360°} = \frac{1}{6}$
9. (a) 359.8 kg   (b) 58.6 kg
10. (b) 112 kg
11.

Method of travel	Angle
Bus	90°
Train	60°
Car	135°
Walk	75°

12. (a) 0 (b) $\frac{3}{5}$ (c) 20   **14.** 72 000
13. (a) 6 (b) 3.55 (c) 3.5   **15.** (a) 275 g

Exam Practice

Non-calculator Paper   Page 112

1. (a) 2609   (b) Sixty million
2. (a) 3   (b)

Monday	⊕ ⊕	
Tuesday	⊕ ⊕ ⊕	⊕ = 4 cars
Wednesday	⊕ ◖	
Thursday	⊕ ⊕ ◿	
Friday	⊕ ⊕ ◖	

3. (a) **7**, **19**, **30**, **105**, **2002**
   (b) (i) 75   (ii) 133   (iii) 286
4. 58 000
5. (a) (i) £2.67   (ii) £2.33
   (b) 1.7 kg   (c) 3500 g
6. (a) Add 6 to the last number, 33.
   (b) Double the last number, 32.
7. (a) (i) 2 (ii) 1 (b) 16 cm (c) 8 cm²
8. (a) $5d$ pence   (b) $(d + 25)$ pence
9. 40
10. (a) 15   (b) 1, 2, 3, 6
11. (a) 10 cm³   (b) $\frac{1}{3}$
12. $P = 26$
13. (a) $\frac{3}{10}$ (b) 70%
14. (a) £37   (b) 15 days
15. (b) $(3, -2)$
16. (a) 59 pence (b) 49 pence (c) 50 pence
17. $-2°C$
18. $a = 99°$
19. (a) 299°   (b) 101 miles
20. (a) (i) 3.28 (ii) 5.4 (iii) 35 (b) 31
21. (a) (i) $7a$ (ii) $11y - x$ (iii) $3a^2$
   (b) (i) $x = 3$ (ii) $x = 4$ (iii) $x = 3$
   (c) 5
22. (a) -8, 16   **23.** (a)
   (b) Negative.
      All even terms
      are negative.

                   (b) $(1, 2)$

24. (a) $\frac{4}{9}$ (b) $\frac{10}{18} = \frac{5}{9}$
  (c) Both have same chance.

25. (a) 34   (b) 25

26. (a) $\angle BAC = 38°$
  (b) **P** and **R**   (SSS or RHS)

27. (a) 256   (b) 60 : 1

28. (a) $5x + 15$ (b) (i) $x = 6$ (ii) $x = 1.5$

29. 8.1 cm²

30. $x = 108°$

31. 5 cm

32. (a)

	1 $\vert$ 8 means 18 km
0	9
1	0 0 1 2 5 7 8
2	0 1 2 5 8 8
3	6

  (b) 18 km

33. (a) $x + (2x - 1) + 3x = 41$
  (b) $x = 7$ Numbers on cards: 7, 13, 21

34. (a) $\frac{1}{10}$ (b) 0.08   (c) 200   (d) $6\frac{4}{15}$

35. $2^3 \times 3^2$

36. (a) 2 : 3   (b) 12π m

37. Greatest: 11.5 kg,   least: 10.5 kg

38. 0924

39. $3n - 1$

40. (a) Enlargement, scale factor $\frac{1}{2}$, centre (0, 1)
  (b) Rotation, 90° anticlockwise, about (1, −1)

41. (a) **R**ed : 0.2,   **B**lue : 0.35,
     Green : 0.15,   **Y**ellow : 0.3
  (b) 35

42. (a) Missing entries are: 1, −1, 1
  (b)

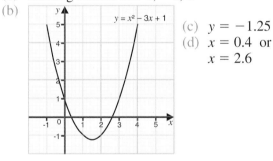

  (c) $y = -1.25$
  (d) $x = 0.4$ or $x = 2.6$

43. $V = 3 \times 4 \times 4 \times 8 = 384$ cm³

44. (a) (i) $5(a - 2)$ (ii) $x(x - 6)$
  (b) (i) $8a + 2$ (ii) $x^2 - 6x + 8$
  (c) $t = \dfrac{W - 3}{5}$ (d) m^5

45. $\sqrt{a^2 + b^2}$ and $4(a + b + c)$
  Both have dimension 1.

46. $1 < x < 2$

Exam Practice

1. (a) −7, 0, 3.5, 5, 10
  (b) 8 thousands, 8000

2. (a)
  (b) Missing entry is: 20
  (c) 27

3. (a) 12 feet (b) 400 metres (c) 200 miles

4.

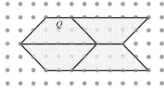

5. (a) **evens**
  (b) **unlikely**

6. (a) **equilateral**   (b) 6
  (c) (i) **trapezium**
     (ii) E.g.

7. Turns through 180° in a clockwise direction.

8. £86

9. (a) $a = 10$ (b) $b = 1.5$

10. (a) 9 years   (b) 18 years

11. (a) $\angle A = 68°$ (b) $x = 124°$

12.

13. $\frac{1}{3}$, 40%, 0.5

14. (a) £22.50
  (b) 80 cards

15. (a) 4 g   (b) (i) $n = 4$ (ii) $m = 3$
  (c) 18   (d) −1

16. (a)   (b) 59 cm²

Not full size

17. (a) 12   (b) 14

18. (a) $\frac{1}{4}$ (b) 800
  (c) 80

19. (a) 3.1   (b) 11.51

20. (a) −1, −5 (b) Subtract 4 from last number.

21. (a) $4m + 5$ (b) $3t - 15$
  (c) (i) $x = \frac{1}{3}$ (ii) $y = 4$ (d) −2

22. (a) 5   (b) $x = 57$

23. (b) $P\,(2.5, 2)$

24. (a) 36   (b) 12.5%

25. (a) If $k = 8$, $\frac{1}{2}k + 1 = 4 + 1 = 5$,
     which is odd.
  (b) If $a = 2$, $b = 3$,   then $2 + 3 = 5$.

26. 52.36 cm²

27. $a = 33°$, $b = 32°$, $c = 65°$

28. (b) Older patients are slower to react.
  (d) 25 years. The point (25, 65) is furthest
            away from the line of best fit.

29. 3.6696 m²

30. (b) 2.15 km,   295°

31. (a) $a = -1$ (b) $m = \frac{2}{3}$ (c) $a = 4$

32. (a) 2.5 kg   (b) 10

33. 128.25 g

34. 2 hours 24 minutes

35. (a) 30   (b) $4 \leqslant t < 6$ (c) 4.13 hours

36. $x = 3.8$

37. 61.1%

38. 2.36

39. (a) −2, −1, 0, 1   (b) $x < \frac{1}{2}$

Index ●●●●●●●●●●●●●●●●●●●●●●●●●●●●●●●